Molecular Biological Markers for Toxicology and Risk Assessment

Molecular Biological Markers for Toxicology and Risk Assessment

Bruce A. Fowler

AMSTERDAM • BOSTON • HEIDELBERG • LONDON • NEW YORK • OXFORD
PARIS • SAN DIEGO • SAN FRANCISCO • SINGAPORE • SYDNEY • TOKYO
Academic Press is an imprint of Elsevier

Academic Press is an imprint of Elsevier
125 London Wall, London EC2Y 5AS, United Kingdom
525 B Street, Suite 1800, San Diego, CA 92101-4495, United States
50 Hampshire Street, 5th Floor, Cambridge, MA 02139, United States
The Boulevard, Langford Lane, Kidlington, Oxford OX5 1GB, UK

Notices
Knowledge and best practice in this field are constantly changing. As new research and experience broaden our understanding, changes in research methods, professional practices, or medical treatment may become necessary.

Practitioners and researchers must always rely on their own experience and knowledge in evaluating and using any information, methods, compounds, or experiments described herein. In using such information or methods they should be mindful of their own safety and the safety of others, including parties for whom they have a professional responsibility.

To the fullest extent of the law, neither the Publisher nor the authors, contributors, or editors, assume any liability for any injury and/or damage to persons or property as a matter of products liability, negligence or otherwise, or from any use or operation of any methods, products, instructions, or ideas contained in the material herein.

Library of Congress Cataloging-in-Publication Data
A catalog record for this book is available from the Library of Congress

British Library Cataloguing-in-Publication Data
A catalogue record for this book is available from the British Library

ISBN: 978-0-12-809589-8

For information on all Academic Press publications
visit our website at https://www.elsevier.com/

Working together
to grow libraries in
developing countries

www.elsevier.com • www.bookaid.org

Typeset by Thomson Digital

Contents

Preface

The field of molecular biomarkers has expanded rapidly in the past several decades due to rapid advances and synergies between a variety of complementary technologies which include advanced analytical instrumentation, robotics, molecular biology, computational modeling, and systems biology. The net result has been the accumulation of a large and growing body of information that could be used for the betterment of public health in the area of chemical/pharmaceutical interactions with biological systems. The "omics" biomarkers (genomics, proteomics, and metabolomics/metabonomics) are current examples of such powerful basic scientific tools, but in order for these technologies to reach their full potential and appreciated value, they must be translated in terms of practical purposes. The field of risk assessment is an obvious choice for such applications but in order for this to occur, there are a number of aspects of the translational process which must be addressed. This book is an attempt to identify and suggest ways in which molecular biomarkers could be translated for risk assessment purposes from a historical perspective. I have attempted to pull insights from existing related fields to help suggest ways to bring forward molecular biomarkers into the mainstream of risk assessment practice. This book is hence intended as an introductory translational text for how existing and evolving molecular biomarkers could be used to inform and help in the formation of better risk assessment decisions from chemical/pharmaceutical exposures. In doing so, the book will emphasize the basic tenets of good science including solid analytical data sets, QA/QC, and biomarker validation studies to help assure the correct prognostic significance of the selected molecular biomarker endpoints in support of risk assessment/risk management decisions based upon sound scientific principles. It is hoped that this text will meet its stated goals and provide some useful guidance to students and others new to this general field and encourage them to move forward with confidence and utilize these tools of modern science.

Bruce A. Fowler, Ph.D., A.T.S.

Presidents Professor of Biomedical Research, University of Alaska- Fairbanks, Fairbanks, Alaska

Adjunct Professor, Rollins School of Public Health, Emory University, Atlanta, Georgia

CHAPTER 1

Molecular Biological Markers for Toxicology and Risk Assessment

1 INTRODUCTION

1.1 Types of Biomarkers (Exposure, Effect Toxicity)

The field of biomarkers has evolved over a number of decades in concert with the number of areas of biomedical science and analytical chemistry. It addresses the need to detect early biological responses at the cellular level which may be correlated with exposures to chemicals, drugs, or mixtures of these agents and that predict health outcomes prior to the onset of metabolic clinical diseases or cancer. These tools are routinely used today for monitoring alterations in organ systems during routine physical examinations or after a clinical event such as a heart attack or acute chemical exposure. Many of the biomarkers in current use have been employed for many years in clinical medicine and their interpretation of predictors of adverse endpoints is now relatively routine based upon decades of experience. However, many of these tests have limited sensitivity or specificity for detecting early signs of cellular damage and only marked changes when extensive organ damage has already occurred. Hence, there is an impetus for development of new, more sensitive, and cost effective tests that will improve the timeliness and precision of decision making processes via incorporation of modern scientific techniques and understanding. The following discussion will briefly review some of the more common clinical biomarker tests which have been used for decades as a way to point out both their useful characteristics and limitations relative to more recent biomarkers such as those based on "omic" technologies. There is a pressing need for validation of biomarker tests at both the cellular and molecular levels of biological organization based upon an integrated and fundamental understanding (Fig. 1.1) of the intracellular mechanisms of toxicity in a target cell population. This is an essential information for biomarker development so that the prognostic implications of any putative biomarker may be correctly interpreted and with regard to mechanisms of cell injury and cell death processes (Fowler,

CONTENTS

1

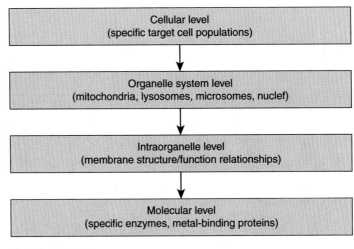

Overall objectives:

1. To understand the relationships which exist between biochemical dysfuction and cellular pathology

2. To elucidate toxic mechanisms by progressive dissection through various levels of intracellular biological organization

FIGURE 1.1 An ultrastructural/biochemical approach initially developed (Fowler, 1980, 1983) for understanding mechanisms of chemical or drug toxicity using the tools of cell biology to dissect down through the various levels of biological organization. This knowledge was further utilized for developing molecular biomarkers of toxicity based upon a basic understanding of which intracellular systems in specific target cell populations were affected by chemical agents.

1987a; Eun et al., 2014) as well as carcinogenesis (Bravaccini et al., 2014; Du et al., 2014; Tabatabaeifar et al., 2014). There are several intracellular validation approaches for validation of molecular biomarkers including correlative cell biology from in vivo animal exposure studies (Fig. 1.2) for toxicity validation purposes at the intracellular levels of biological organization (Fowler, 1980). But more recently studies using in vitro systems coupled with computational systems biology methods (Fowler, 2013) have largely replaced the use of intact animals for biomarker development (Fig. 1.3). These newer approaches will be discussed in greater detail in subsequent chapters in relation to translation of molecular biomarkers for risk assessment practice. As noted the field of biomarkers is rapidly evolving, but the basic scientific principles, given later, that were articulated by Bradford Hill (1965) with regard to linkages between chemical exposures, biological outcomes, and causality remain as critical elements.

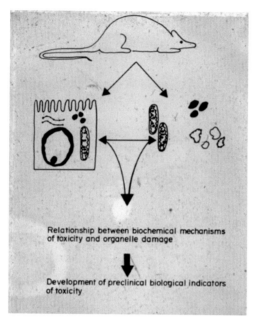

FIGURE 1.2 Conceptual diagram for cell biology approach to biomarker (preclinical indicator) development and validation using intact animal model systems and in vivo exposures. *From Fowler (1980), with permission from Elsevier.*

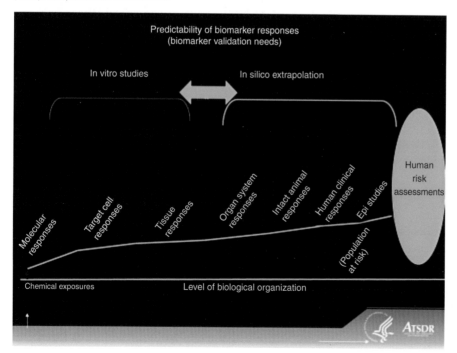

FIGURE 1.3 Conceptual diagram for development of biomarkers across different levels of biological organization using data from in vitro systems coupled with computer-assisted extrapolations to provide information for risk assessment linkages to human populations.

Bradford Hill's principles of causality in clinical research

- Strength of the association
- Specificity of the association
- Dose-response evidence
- Biological plausibility of the hypothesis
- Coherence of the evidence
- Temporality
- Consistency of results across studies

In addition, major interagency initiatives, such as the USEPA NEXGEN (Krewski et al., 2014) and Tox21 programs (Thomas et al., 2013), are attempting to incorporate molecular biomarker endpoints into regulatory decision-making processes regarding individual chemical exposure (particularly at low dose levels) and mixture exposure. This approach offers a great potential for improving the precision of such important risk assessments. Incorporation of genomic profiling derived from the NHANES database has permitted more specific risk assessments for low-dose lead exposures in the general US population. This database has also been used to document evidence that molecular biomarkers (ALAD) may be further fine-tuned to identify sensitive subpopulations on the basis of genetic inheritance (Scinicariello et al., 2010). Similar approaches for other chemicals should permit risk assessors to address more precisely if a given chemical or mixture exposure is more harmful to a specific group of individuals. This book briefly reviews the history of biomarkers over the last 50 years and discusses the evolution of some of these powerful tools. Specific examples of how they have been applied to produce better science-based public health decisions are given. Included is a forward looking discussion of the current status of biomarkers and how they could be applied more effectively for risk assessments linked to human epidemiological studies via computer modeling techniques.

2 ENZYME ACTIVITY–BASED BIOMARKERS

Measurement of enzyme activities, such as that of lactic dehydrogenase (LDH), and a number of transaminase enzymes found in serum including alanine transaminase and serum glutamate oxaloacetic transaminase, have been used effectively over a long period in clinical settings to follow degrees of tissue damage in the liver from chemical exposures and in the heart from cardiovascular events (Ozer et al., 2008).

LDH is primarily a cytosolic-based enzyme that has been shown to leak from damaged cells and is readily measurable in accessible body fluids such as serum. The increased presence of this enzyme activity is indicative of widespread cell injury. Alanine transaminase and serum glutamate oxaloacetic transaminase activities have shown similar utility in assessing liver damage from chemical exposures. Elevated levels of these enzyme activities in the serum are used as indicators of the need for further follow-up evaluations. These biomarkers provide rapid, cost-effective information that guide the need for more extensive and costly diagnostic studies. The ability to effectively "triage" is an important attribute (Ozer et al., 2008). Measurements of these enzyme activities have been used clinically to good effect for many years; however, newer biomarkers, discussed in succeeding sections, overcome some of their limitations due to low sensitivity.

2.1 Tissue-Specific Isozymes

Tissue-specific isozymes for a number of these enzyme activities permit delineation of the organ sources of the enzyme and hence the sites of damage. These are clearly potentially very useful and more specific diagnostic biomarker tools. Sensitivity is a major issue with measuring enzyme activities since many assays do not show significant changes until extensive tissue damage has occurred. Ideally, one would like to detect ongoing cellular damage at the earliest stage so that effective interventions could be initiated. Correlative histopathological and ultrastructural studies are very important for interpretation of biomarker data in order to delineate *which* cells within an affected organ are being damaged by a toxic agent and hence the origin of the biomarker response is known. Usually only a subset of cells within a target organ show damage from a given toxic agent at lower dose levels, and these effects may be missed if the measurement of the selected biomarker is not sufficiently sensitive.

3 CURRENTLY MEASURED BIOMARKER PROTEINS

A number of specific proteins have been measured in accessible matrices such as blood and urine. These have been and are currently used as biomarkers for assessing metabolic diseases and effects of chemicals and pharmaceuticals on common target organ systems, such as the hematopoietic system or kidneys and liver. The measurements of proteins offer improved sensitivity and specificity over enzyme activity measurements since they can be measured by immune assay techniques and can be automated for increased speed and cost efficiencies. The following represent a short list of the more commonly measured proteins that are clinically useful.

3.1 Glycosylated Hemoglobin

Glycosylated hemoglobin (HbA1c) is used as a biomarker for diabetes mellitus. These measurements have been used effectively for several decades as a diagnostic indicator for this common disease. Nevertheless, care must be taken with interpretation of findings since ingestion of aspirin may also increase hemoglobin acetylation and lead to misinterpretation of results unless further analytical studies are also performed (Nathan, 1983).

3.2   Prostate-Specific Antigen

This protein is routinely used as a marker of male prostate hyperplasia and cancer but may be altered by infections; so findings of increases in blood levels of this protein must be interpreted with caution and other measures incorporated. Nonetheless, the prostate-specific antigen (PSA) test is widely accepted as a screening tool for prostate cancer. More recently, the prognostic value of the PSA test has been shown to be enhanced by combined measurement of serum LDH activity (Kelly et al., 1993; Taplin et al., 2005).

3.3   Beta 2 Microglobulin

Beta 2 microglobulin (B2M) low molecular weight protein (11,800 Da) is found on the surface of nucleated white cells has been used for several decades as a prognostic marker of cancers of the immune system when measured in blood (Greipp et al., 1993) and also as biomarker of low molecular weight tubular proteinuria observed in persons with elevated exposures to metals such as cadmium (Nordberg et al., 2015).

3.4   Retinol-Binding Protein

Retinol-binding proteins (RBPs) are another low molecular weight protein marker with a mass of 21 kDa which transport retinoic acid in blood. As a low molecular weight protein family, they are readily filtered by the glomerulus and normally reabsorbed by the kidney proximal tubule. These proteins may also appear in the urine as a tubular proteinuria produced by elevated exposures to metals such as cadmium (Bernard et al., 1997).

3.5   Alpha-1 Microglobulin

Alpha-1 microglobulin (A-1MG) protein has molecular mass of 26,000 Da and has been reported to be increased in urine samples of persons exposed to arsenic in drinking water in China (Feng et al., 2013) and thus reported as a measure of renal toxicity.

3.6   Stability Issues With B2M, RBP, and A-1MG

Studies (Donaldson et al., 1989) have reported that B2M, RBP and A-1MG decreased stability in urine samples with pH values below 7.0. The problem was most severe for B2M. These data illustrate the general importance of conducting stability studies for all biomarker endpoints measured in biological matrices, such as urine, so that findings may be correctly interpreted. This is of particular importance for samples collected during field studies in remote areas where access to refrigeration may be limited.

3.7 Cystatin c

Cystatin c is a protein with a molecular mass of 13,300 Da which is readily filtered by the glomeruli and appears in the urine in cases of kidney injury. It is a cysteine proteinase inhibitor (Barrett et al., 1984). Metaanalysis studies have demonstrated that it is another effective biomarker of kidney tubular injury (Dharnidharka et al., 2002).

3.8 Kidney Injury Molecule-1

Kidney injury molecule-1 is a kidney tubule cell-specific protein of intermediate molecular weight which is released into the urines of persons with increased exposures to toxic metals from damaged kidney tubule cells (Bailly et al., 2002; Han et al., 2002; van Timmeren et al., 2007; Prozialeck et al., 2009a,b; Vaidya et al., 2009). The major advantage of measuring this protein in urine is that it is a component of the kidney proximal tubule membrane and hence a more specific measure of damage to the kidney proximal tubule cells in contrast with proteins produced in other tissues and not reabsorbed due to loss of proximal tubule cell functionality.

3.9 Metallothionein

Metallothionein is a small inducible, cysteine-rich protein which binds a number of toxic and essential metals and appears to be capable of playing a number of roles in the biology of cells. It exists as a number of isoforms and members of this family are found across a number of phyla (Fowler, 1987b; Fowler et al., 1987). It is thought to be a useful general biomarker of susceptibility to toxicity for metals such as cadmium (Fowler, 2009).

3.10 Lead-Binding Proteins

ALAD and low molecular weight lead-binding proteins appear to play major roles in the bioavailability of lead to sensitive molecular sites primarily involved in mediating toxicity. ALAD is the major transporter of lead in blood (Scinicariello et al., 2007; Tian et al., 2013). The lower molecular weight lead-binding proteins (PbBPs) appear to play major roles in the intracellular compartmentalization of lead in sensitive target organs such as the kidney and brain (Oskarsson et al., 1982; Fowler et al., 1993; Quintanilla-Vega et al., 1995; Smith et al., 1998) at low, environmental dose levels of lead. These molecules hence play major regulatory roles in mechanisms of lead toxicity and the determination of thresholds of toxicity for risk assessment purposes by regulating the intracellular bioavailability of lead to sensitive target sites. In recent years, the application of public health genomics has helped to further elucidate the importance of ALAD polymorphisms in mediating genetic-based differences in susceptibility to lead-induced cardiovascular effects (Scinicariello et al., 2010).

4   OMIC BIOMARKERS

Omic biomarkers (genomics, proteomics, metabolomics/metabonomics) are the most recently evolved classes of biomarkers and are receiving a great deal of attention due, in part, to the insights they can provide into the fundamental processes of cell injury and cell death. These classes of biomarkers will be examined in greater detail in subsequent chapters but a brief overview is given in following sections.

4.1   Genomics

Genomics or the study of altered responses of genes to chemical or physical stimuli, such as ionizing radiation (Lee et al., 2014), is well developed, and the techniques used can detect biological responses of cells at the level of the genome. The value of such data rests with a better understanding of which gene systems (eg, oncogenes) are reacting to chemical or physical stimuli. This, in turn, can help elucidate causal pathways for early detection of adverse health interactions with epigenetic regulatory systems (Meaney and Szyf, 2005) as well as for predicting the course of diseases such as TB or latent TB infections (Wu et al., 2014) or various cancers (Catto et al., 2011; Du et al., 2014) or drug/chemical toxicities (Eun et al., 2014). Information about epigenetic factors is hence vital to accurate interpretation of genomic responses for risk assessment purposes and predicting health outcomes (Simon, 2011). On the other hand, genomic approaches have been usefully applied to various cancers (Bravaccini et al., 2014; Jewell et al., 2014; Singh et al., 2014; Tabatabaeifar et al., 2014), reproductive disorders (Egea et al., 2014), and early responses to toxic chemicals (Fowler and Akkerman, 1992) and pharmaceuticals (Dalabira et al., 2014; Padulles et al., 2014). Such data are clearly important for identifying sensitive subpopulations for risk assessment purposes. As with the other omic biomarker categories discussed later, these technologies generate enormous quantities of data which require development of computerized data analysis systems that incorporate other biological data. These sophisticated systems are essential in order to evaluate information placed in such databases (Izzo et al., 2014; Renaud et al., 2014; Zlobec et al., 2014) and permit systems biology approaches to the data for risk assessment purposes. See Fowler (2013) for a more general discussion on this topic. The data generated from systems biology approaches may hence provide useful information on *which* molecular pathways are responding to stimuli of interest and potentially provide bridges to understanding activation of adverse outcome pathways such as the apoptotic cascades. These mechanisms are discussed further in subsequent chapters in relation to translation of molecular biomarkers for risk assessment practice.

4.2 Proteomics (2-D Gels, HPLC–MS, ICP-MS, Posttranslational Modifications, Epigenetics)

Proteomics is essentially the evaluation of alterations of expressed proteins or gene products in response to chemical or physical stimuli. Two-dimensional gel electrophoresis (2-DGE) was the first approach used to permit appreciation of the complexity of altered protein expression patterns in response to these stimuli from exposures to agents, such as arsenic, as a function of dose and time. These tools have proven to be very useful in mixture exposure situations and in distinguishing the relative toxicity of two similar semiconductor compounds (Fowler et al., 2005, 2008). In more recent years, 2-DGE of proteins has been expanded by development of silver staining methods for proteins and development of digitizing computer image analytical programs (Stessl et al., 2009; Minden, 2012). In addition, HPLC/ICP/MS approaches coupled with computerized analysis programs (Bouatra et al., 2013) have greatly speeded up the analyses, improved interpretation of 2-DGE data, and reduced costs of applying this technology for biomarker development and data interpretation as discussed later. Urinary proteomics has been applied as a biomarker approach for assessment of chronic kidney disease and also to follow renal disease progression (Schanstra and Mischak, 2015). Comparative genomic and proteomic studies (Nicolini and Pechkova, 2010; Goff et al., 2013; Eun et al., 2014) have demonstrated the value of both approaches in delineating specific patterns of drug-induced liver injury in rats. More recent advances in nanoproteomics based on microfluidic systems (Nicolini and Pechkova, 2010; Ray et al., 2011; Yang et al., 2014) coupled with onboard computer operating and analysis systems now permit analyses of protein expression patterns in small biological samples. These advances in technology have clearly opened up new possibilities for proteomic biomarker development.

The advent of these sensitive electrophoretic and mass spectrometry techniques has permitted evaluation of a number of proteins as possible biomarkers for early detection of tissue injury from exposure to chemicals and drugs. These biological endpoints offer low cost, increased sensitivity, and quick information for early detection of cellular responses to chemical- or drug-induced insults. A major remaining issue with these biomarker data is interpretation of their prognostic significance. Are changes in these endpoints harbingers of eventual cell death or cancer or simply a reversible cellular "sneezes" which are corrected by other cellular machinery such as the "stress protein response"?

This approach has been used to distinguish gender differences in responses by using age- and race-matched renal tubule cells exposed to semiconductor elements in vitro at equivalent concentrations (Fowler et al., 2008). This subject area will be discussed in greater detail in subsequent chapters with regard to "adverse outcome pathways" and the EPA NEXGEN and Tox21 initiatives for risk assessment.

4.3 Metabolomics/Metabonomics (Porphyrins, Metabolites, Protein Fragments)

Metabolomics/metabonomics approaches utilize sophisticated analytical methods such as HPLC/ICP/MS alone or in combination with computerized analysis programs to evaluate alterations in metabolic products produced by chemical or physical disruption of metabolic pathways. For example, chemical- or drug-induced disturbances in the heme biosynthetic pathway with resultant increases in the urinary excretion of specific porphyrins due to inhibition of that essential cellular pathway at specific enzymatic steps has proven useful in diagnosing degrees of organ damage (Garcia-Vargas et al., 1994) and cancer (Chan et al., 2014) in humans. The linkage of such data with morphological finding have proven useful (Fowler et al., 2008) for validation of the porphyrin profile data for risk assessment purposes. Since the heme biosynthetic pathway is essential for life and highly conserved across species, the data from these biomarker studies in experimental animal studies may be readily applied for risk assessments in humans and other species with adjustments for chemical dosimetry. In addition, the advent of modern HPLC/ICP/MS systems has permitted the exploration of this approach to other metabolic pathways (Li et al., 2014; Ramautar et al., 2014), but no new scientific concepts are involved. The application of these sophisticated analytical approaches to cellular metabolites and protein fragments has engendered the growing field of metabolomics/metabonomics. This potentially provides additional internal biological information with regard to ongoing processes, such as oxidative stress, obesity (Rauschert et al., 2014), and the roles of mitochondria (Demine et al., 2014), in response to chemical or drug insults. Such information is potentially useful for risk assessments of the degree of ongoing tissue damage or recovery following cessation of exposures or the effectiveness of therapeutic treatment regimens. It is clear from the previous discussions that each of these omic approaches generate enormous quantities of data regarding different biological systems on individual chemicals or drugs alone or in combination. Effective evaluation of such large and complex data sets requires application of new and innovative computational methodologies including artificial intelligence programs as previously discussed (Fowler, 2013). Only when all available data are analyzed as a whole can existing knowledge be optimized for risk assessment.

5   COMPUTATIONAL TOXICOLOGY APPROACHES

In addition to the laboratory-based approaches noted earlier, it is also possible to take advantage of published findings for biomarker development through the use of computer-assisted data mining approaches (Ruiz et al., 2010a,b, 2015) to evaluate information in public databases, such as NHANES. Information extracted from these databases can be used to develop both pathway and network knowledge maps (Ruiz et al., 2015) which can shed light on likely biological endpoints for biomarker development and linkages between these sites and sites related to cell death and replacement (eg, adverse outcome pathways). In addition, rigorous evaluation of the peer-reviewed published literature permit metaanalyses of the data which are valuable in confirming findings and supporting risk assessment conclusions. Development of knowledge maps from computerized analyses of the published literature may also aid in the development of testable hypotheses by permitting a more global view of how chemicals may interact with different genomic, proteomic, and metabolic systems and thus guide and expedite laboratory-based research in more efficient manner. (See Fowler, 2013 for discussion.)

5.1   Data Mining of Public Databases

There are large quantities of high quality data located in both public (eg, CDC-NHANES, NIH PUBMED) and private (eg, Google Scholar, Big Pharma) databases. See Fowler (2013) and Ruiz et al. (2015) for a recent review. The value of these data is that they can frequently be accessed and analyzed in a timely and cost-effective manner.

5.2   Metaanalysis Approaches

Metaanalysis procedures have evolved extensively over the past 20 years and provide a means for grouping and comparing the results of published studies to delineate overall conclusions about associations between exposures and health outcome measures. These data mining approaches and techniques may expedite the development of testable hypotheses and stimulate mechanistic research into the roles of genomic alleles in mediating susceptibility to toxic agents such as lead (Scinicariello et al., 2007, 2010).

5.3   Network and Pathway Analyses and Systems Biology Approaches Using Data From the Peer-Reviewed Published Literature

On another level of biological organization, data mining of the published literature coupled with computerized analyses can produce knowledge maps (Ruiz et al., 2015) that include both biochemical network and pathway analyses which can also be very instructive in supporting specific hypotheses and laboratory-based experiments. These systems biology approaches using computer modeling techniques to develop testable hypotheses also clearly expedite laboratory-based research and potentially reduce the cost of research.

All of the previously mentioned approaches can provide insights that may expedite mechanistic biomarker-based research that, in turn, can delineate adverse outcome pathway mechanisms of toxicity.

A major value of computational modeling in simulating needed experiments in silico rests with speed and reduced costs. While these simulations are not an actual substitute for wet laboratory evaluations, these modeling approaches have proven to be useful in delineating a "ball park" answer which has proven quite useful in making good decisions in emergency situations such as the Deepwater Horizon Gulf oil spill (ATSDR, 2011).

6   APPLICATION OF MOLECULAR BIOMARKERS FOR RISK ASSESSMENT

The previous discussion is a brief overview of the major classes of biomarkers and ways in which these molecular tools have been and are continuing to be used to provide information about the ways in which chemicals, drugs, and physical entities may influence biological systems under both acute and chronic exposure conditions. The challenge now is to translate this body of useful basic scientific information into practical and credible risk assessment–based decision making. The overall long term goal would be to replace proscribed uncertainty factors of 3 or 10 with more precise estimates of risk based on actual mechanistic scientific data. A fair question is whether this goal can, in fact, be accomplished. Given the overview of the accelerating progress made to date on understanding the relationships between the, various classes of biomarkers and mechanisms of cell injury/cell death and cancer, the answer to this question is positive but it will take hard work and commitment. Chapter 5 briefly examines currently known linkages between cellular processes which support life and those processes (eg, apoptosis) which lead to cellular death or cancer. Ultimately the fundamental question, of why do cells die (see Fowler, 1987a for a discussion) and how can the tools of modern science provide useful information to follow the underlying processes via use of molecular biomarker endpoints, will be revisited. Examples of how translation of this basic knowledge could be used to prevent or interdict adverse health outcomes by low-dose exposures to common toxic agents such as arsenic and cadmium on an individual or mixture basis will be discussed as case studies.

A final section will deal with risk communication techniques, such as information mapping, that are capable of translating some of the complex technical information derived from omic and computational approaches into a teaching format of informational value to persons with more limited technical backgrounds and who need to understand the information in order to make sound societal public health decisions.

References

ATSDR, 2011. Sharing our stories: the deepwater horizon oil spill. ATSDR/CDC website, Atlanta.

Bailly, V., et al., 2002. Shedding of kidney injury molecule-1, a putative adhesion protein involved in renal regeneration. J. Biol. Chem. 277 (42), 39739–39748.

Barrett, A.J., et al., 1984. The place of human gamma-trace (cystatin C) amongst the cysteine proteinase inhibitors. Biochem. Biophys. Res. Commun. 120 (2), 631–636.

Bernard, A., et al., 1997. Urinary biomarkers to detect significant effects of environmental and occupational exposure to nephrotoxins. IV. Current information on interpreting the health implications of tests. Ren. Fail. 19 (4), 553–566.

Bouatra, S., Aziat, F., Mandal, R., Guo, A.C., Wilson, M.R., Knox, C., et al., 2013. The human urine metabolome. PloS One 8 (9), e73076.

Bravaccini, S., et al., 2014. New biomarkers to predict the evolution of in situ breast cancers. Biomed. Res. Int. 2014, 159765.

Catto, J.W., et al., 2011. MicroRNA in prostate, bladder, and kidney cancer: a systematic review. Eur. Urol. 59 (5), 671–681.

Chan, A.W., et al., 2014. Potential role of metabolomics in diagnosis and surveillance of gastric cancer. World J. Gastroenterol. 20 (36), 12874–12882.

Dalabira, E., et al., 2014. DruGeVar: an online resource triangulating drugs with genes and genomic biomarkers for clinical pharmacogenomics. Public Health Genomics 17, 265–271.

Demine, S., et al., 2014. Unraveling biochemical pathways affected by mitochondrial dysfunctions using metabolomic approaches. Metabolites 4 (3), 831–878.

Dharnidharka, V.R., et al., 2002. Serum cystatin C is superior to serum creatinine as a marker of kidney function: a meta-analysis. Am. J. Kidney Dis. 40 (2), 221–226.

Donaldson, M.D., et al., 1989. Stability of alpha 1-microglobulin, beta 2-microglobulin and retinol binding protein in urine. Clin. Chim. Acta 179 (1), 73–77.

Du, M., et al., 2014. Clinical potential role of circulating microRNAs in early diagnosis of colorectal cancer patients. Carcinogenesis 35, 2723–2730.

Egea, R.R., et al., 2014. OMICS: current and future perspectives in reproductive medicine and technology. J. Hum. Reprod. Sci. 7 (2), 73–92.

Eun, J.W., et al., 2014. Characteristic molecular and proteomic signatures of drug-induced liver injury in a rat model. J. Appl. Toxicol. 35, 152–164.

Feng, H., et al., 2013. Biomarkers of renal toxicity caused by exposure to arsenic in drinking water. Environ. Toxicol. Pharmacol. 35 (3), 495–501.

Fowler, B.A., 1980. Ultrastructural morphometric/biochemical assessment of cellular toxicity. In: Witschi, H.P. (Ed.), Scientific Basis of Toxicity Assessment. Elsevier, Amsterdam, The Netherlands, pp. 211–218.

Fowler, B.A., 1983. The role of ultrastructural techniques in understanding mechanisms of metal-induced nephrotoxicity. Fed. Proc. 42, 2957–2964.

Fowler, B.A., 1987a. Mechanisms of cell injury: implications for human health. Dahlem Workshop ReportsJohn Wiley & Sons, Hoboken, NJ.

Fowler, B.A., 1987b. Intracellular compartmentation of metals in aquatic organisms: relationships to mechanisms of cell injury. Environ. Health Perspect. 71, 121–128.

Fowler, B.A., 2009. Monitoring of human populations for early markers of cadmium toxicity: a review. Toxicol. Appl. Pharmacol. 238 (3), 294–300.

Fowler, B.A., 2013. Computational Toxicology: Methods and Applications for Risk Assessment. Elsevier, Amsterdam, The Netherlands, 258 pp.

Fowler, B.A., Akkerman, M., 1992. The role of Ca^{2+} in cadmium-induced renal tubular cell injury. IARC Sci. Publ. 118, 271–277.

Fowler, B.A., Hildebrand, C.E., Kojima, Y., Webb, M., 1987. Nomenclature of metallothionein. Metallothionein IIBirkhäuser, Basel, Switzerland, pp. 19–22.

Fowler, B.A., et al., 1993. Implications of lead binding proteins for risk assessment of lead exposure. J. Expo. Anal. Environ. Epidemiol. 3 (4), 441–448.

Fowler, B.A., et al., 2005. Metabolomic and proteomic biomarkers for III-V semiconductors: chemical-specific porphyrinurias and proteinurias. Toxicol. Appl. Pharmacol. 206 (2), 121–130.

Fowler, B.A., et al., 2008. Proteomic and metabolomic biomarkers for III-V semiconductors: and prospects for application to nano-materials. Toxicol. Appl. Pharmacol. 233 (1), 110–115.

Garcia-Vargas, G.G., et al., 1994. Altered urinary porphyrin excretion in a human population chronically exposed to arsenic in Mexico. Hum. Exp. Toxicol. 13 (12), 839–847.

Goff, D.J., et al., 2013. A Pan-BCL2 inhibitor renders bone-marrow-resident human leukemia stem cells sensitive to tyrosine kinase inhibition. Cell Stem Cell 12 (3), 316–328.

Greipp, P.R., et al., 1993. Plasma cell labeling index and beta 2-microglobulin predict survival independent of thymidine kinase and C-reactive protein in multiple myeloma. Blood 81 (12), 3382–3387.

Han, W.K., et al., 2002. Kidney injury molecule-1 (KIM-1): a novel biomarker for human renal proximal tubule injury. Kidney Int. 62 (1), 237–244.

Hill, A.B., 1965. The environment and disease: association or causation? Proc. R. Soc. Med. 58 (5), 295–300.

Izzo, M., et al., 2014. A digital repository with an extensible data model for biobanking and genomic analysis management. BMC Genomics 15 (Suppl. 3), S3.

Jewell, R., et al., 2014. The clinicopathological and gene expression patterns associated with ulceration of primary melanoma. Pigment Cell Melanoma Res. 28, 94–104.

Kelly, W.K., et al., 1993. Prostate-specific antigen as a measure of disease outcome in metastatic hormone-refractory prostate cancer. J. Clin. Oncol. 11 (4), 607–615.

Krewski, D., et al., 2014. A framework for the next generation of risk science. Environ. Health Perspect. 122 (8), 796–805.

Lee, K.F., et al., 2014. Gene expression profiling of biological pathway alterations by radiation exposure. Biomed. Res. Int. 2014, 834087.

Li, J., et al., 2014. Metabolic profiling study on potential toxicity and immunotoxicity-biomarker discovery in rats treated with cyclophosphamide using HPLC-ESI-IT-TOF-MS. Biomed. Chromatogr. 29, 768–776.

Meaney, M.J., Szyf, M., 2005. Environmental programming of stress responses through DNA methylation: life at the interface between a dynamic environment and a fixed genome. Dialogues Clin. Neurosci. 7 (2), 103–123.

Minden, J.S., 2012. Two-dimensional difference gel electrophoresis. Methods Mol. Biol. 869, 287–304.

Nathan, D.M., Francis, T.B., Palmer, J.L., 1983. Effect of aspirin on determination of glycosylated hemoglobin. Clin. Chem. 29 (3), 466–469.

Nicolini, C., Pechkova, E., 2010. Nanoproteomics for nanomedicine. Nanomedicine 5 (5), 677–682.

Nordberg, G., et al., 2015. Cadmium. In: Nordberg, G. et al., (Ed.), Handbook on Toxicology of Metals. Elsevier, Amsterdam, The Netherlands, pp. 667–716.

Oskarsson, A., et al., 1982. Intracellular binding of lead in the kidney: the partial isolation and characterization of postmitochondrial lead binding components. Biochem. Biophys. Res. Commun. 104 (1), 290–298.

Ozer, J., et al., 2008. The current state of serum biomarkers of hepatotoxicity. Toxicology 245 (3), 194–205.

Padulles, A., et al., 2014. Developments in renal pharmacogenomics and applications in chronic kidney disease. Pharmgenomics Pers. Med. 7, 251–266.

Prozialeck, W.C., et al., 2009a. Expression of kidney injury molecule-1 (Kim-1) in relation to necrosis and apoptosis during the early stages of Cd-induced proximal tubule injury. Toxicol. Appl. Pharmacol. 238 (3), 306–314.

Prozialeck, W.C., et al., 2009b. Preclinical evaluation of novel urinary biomarkers of cadmium nephrotoxicity. Toxicol. Appl. Pharmacol. 238 (3), 301–305.

Quintanilla-Vega, B., et al., 1995. Lead-binding proteins in brain tissue of environmentally lead-exposed humans. Chem. Biol. Interact. 98 (3), 193–209.

Ramautar, R., et al., 2014. CE-MS for metabolomics: developments and applications in the period 2012–2014. Electrophoresis 36, 212–224.

Rauschert, S., et al., 2014. Metabolomic biomarkers for obesity in humans: a short review. Ann. Nutr. Metab. 64 (3–4), 314–324.

Ray, S., et al., 2011. Emerging nanoproteomics approaches for disease biomarker detection: a current perspective. J. Proteomics 74 (12), 2660–2681.

Renaud, G., et al., 2014. trieFinder: an efficient program for annotating digital gene expression (DGE) tags. BMC Bioinformatics 15 (1), 329.

Ruiz, P., et al., 2010a. Physiologically based pharmacokinetic (PBPK) tool kit for environmental pollutants—metals. SAR QSAR Environ. Res. 21 (7–8), 603–618.

Ruiz, P., et al., 2010b. Interpreting NHANES biomonitoring data, cadmium. Toxicol. Lett. 198 (1), 44–48.

Ruiz, P., et al., 2015. Pathway analysis of different persistent organic pollutants suggest common disease connections. Society of Toxicology Meeting, San Diego.

Schanstra, J.P., Mischak, H., 2015. Proteomic urinary biomarker approach in renal disease: from discovery to implementation. Pediatr. Nephrol. 30, 713–725.

Scinicariello, F., et al., 2007. Lead and delta-aminolevulinic acid dehydratase polymorphism: where does it lead? A meta-analysis. Environ. Health Perspect. 115 (1), 35–41.

Scinicariello, F., et al., 2010. Modification by ALAD of the association between blood lead and blood pressure in the U.S. population: results from the Third National Health and Nutrition Examination Survey. Environ. Health Perspect. 118 (2), 259–264.

Simon, R., 2011. Genomic biomarkers in predictive medicine: an interim analysis. EMBO Mol. Med. 3 (8), 429–435.

Singh, A.K., et al., 2014. Human mitochondrial genome flaws and risk of cancer. Mitochondrial DNA 25 (5), 329–334.

Smith, D.R., et al., 1998. High-affinity renal lead-binding proteins in environmentally-exposed humans. Chem. Biol. Interact. 115 (1), 39–52.

Stessl, M., et al., 2009. Influence of image-analysis software on quantitation of two-dimensional gel electrophoresis data. Electrophoresis 30 (2), 325–328.

Tabatabaeifar, S., et al., 2014. Use of next generation sequencing in head and neck squamous cell carcinomas: a review. Oral Oncol. 50, 1035–1040.

Taplin, M.E., et al., 2005. Prognostic significance of plasma chromogranin a levels in patients with hormone-refractory prostate cancer treated in Cancer and Leukemia Group B 9480 study. Urology 66 (2), 386–391.

Thomas, R.S., et al., 2013. Incorporating new technologies into toxicity testing and risk assessment: moving from 21st century vision to a data-driven framework. Toxicol. Sci. 136 (1), 4–18.

Tian, L., et al., 2013. Lead concentration in plasma as a biomarker of exposure and risk, and modification of toxicity by delta-aminolevulinic acid dehydratase gene polymorphism. Toxicol. Lett. 221 (2), 102–109.

Vaidya, V.S., et al., 2009. A rapid urine test for early detection of kidney injury. Kidney Int. 76 (1), 108–114.

van Timmeren, M.M., et al., 2007. Tubular kidney injury molecule-1 (KIM-1) in human renal disease. J. Pathol. 212 (2), 209–217.

Wu, L.S., et al., 2014. Systematic expression profiling analysis identifies specific microRNA-gene interactions that may differentiate between active and latent tuberculosis infection. Biomed. Res. Int. 2014, 895179.

Yang, X., et al., 2014. Analysis of the human urine endogenous peptides by nanoparticle extraction and mass spectrometry identification. Anal. Chim. Acta 829, 40–47.

Zlobec, I., et al., 2014. A next-generation tissue microarray (ngTMA) protocol for biomarker studies. J. Vis. Exp. 91, 51893.

CHAPTER 2

Historical Development of Biomarkers

1 BIOMARKERS OF EXPOSURE

1.1 Introduction

In order to use molecular biomarkers for improving risk assessments, there are several components of scientific information which are essential. These are (1) biomarkers of exposure, (2) biomarkers of effect, and (3) biomarkers of toxicity or adverse outcomes (Fowler, 2012). Ultimately, it is desirable to have information on all three of these elements in order to inform and improve risk assessments with the application and integration of modern scientific knowledge to produce a more comprehensive understanding of risks from chemical or drug exposures. This is rather like generating a three-leg scientific stool utilizing data from all three of these major areas to inform a more precise picture of risk. The rapidly evolving tools of computational toxicology could then be used to integrate and analyze the large combined dataset. This chapter will focus on a discussion of biomarkers of exposure which is the first element in this process and will take up an overview of how data generated by the new tools of analytical chemistry coupled with computational modeling approaches can greatly enhance risk assessments. This information will be integrated with information on biomarkers of effect and toxicity/adverse outcomes in subsequent chapters. Translation of these biomarker data and research needs, so that this body of information may be utilized in needed risk assessments, will be taken up in penultimate chapters. Methods for communication of risk outcomes from complex biomarker data sets (eg, information mapping technology) into language of value to societal decision makers will be discussed later in this book.

2 EXPOSOME AND BIOMARKERS OF EXPOSURE

In general terms, the term "exposome" represents the totality of all the chemical and physical factors to which humans or other species of interest are exposed (Rappaport, 2011) but, in addition, speaking more broadly, includes how a given individual metabolizes a chemical or mixture of chemicals to generate metabolites (Wild et al., 2013). This situation clearly includes a very

CONTENTS

17

Molecular Biological Markers for Toxicology and Risk Assessment. http://dx.doi.org/10.1016/B978-0-12-809589-8.00002-0

large and growing list of chemical agents and metabolites which represent an enormous challenge to the risk assessment community. The issue of chemical mixture exposures and the growing number of nanomaterials will be taken up separately in subsequent chapters since mixture exposures are the most common scenario and a major challenge to risk assessors.

Advances in analytical methods have greatly increased the ability of chemists to accurately measure wide variety of chemicals and pharmaceutical agents and their metabolites. Outstanding major issues are centered around interpretation of the analytical data in relation to health outcomes and ascertaining possible interactions among the measured chemical entities. This is clearly not a simple task but integration of analytical data with information on genetic inheritance, biomarkers of effect, and adverse outcomes via computational technologies is an increasingly viable approach that should lead to increased risk assessment precision and ultimately individualized risk assessments in the long term.

3   SPECIFIC BIOMONITORING ANALYTICAL METHODOLOGIES

The following discussion will briefly review some of the more common analytical methodologies and highlight the ways in which these tools have contributed toward improvements in biomonitoring for exposure assessments and ultimately risk assessment practice.

3.1   High Performance Liquid Chromatography

High performance liquid chromatography (HPLC) has evolved into a major analytical workhorse for a variety of chemical agents and their metabolites since its introduction approximately 40 years ago. The widespread use of this methodology has been stimulated by number of factors including relatively low cost, ease of use, and the development of onboard computer systems which permit menu driven analyses and the evolution of various types of columns including reverse phase, ion exchange, and size exclusion (Gallagher et al., 2014; Pedrali et al., 2014; Qiao et al., 2014; Song et al., 2014). More recently, microfluidic column technology Li et al., 2013; Luong et al., 2013; Nie and Kennedy, 2013; Thurmann and Belder, 2014; Yang et al., 2014a; Gallagher et al., 2014) has greatly reduced sample size requirements. Improvements in HPLC detectors and coupling of these instruments to other highly sensitive analytical systems such as mass spectrometers have further expanded the range and sensitivity of biological monitoring studies. These technical improvements coupled with the evolution of QA/QC protocols and interlaboratory comparison studies have contributed greatly to the acceptance of chemical biomonitoring data as important and reliable exposure components for risk assessment practice (Gambrill and Shlonsky, 2000; Krewski et al., 2014). This technology has been applied in studies of arsenicals

(Amayo et al., 2014; Ukena et al., 2014; Wang et al., 2014) and a host of organic chemicals and pharmaceutical agents (Jain and Wang, 2011; Miller et al., 2014; Naldi et al., 2014; Patterson et al., 2009; Shi et al., 2014).

3.2 Gas Chromatography

Gas chromatography is another column-based chromatographic technique which separates organic chemicals or organometallic chemicals by retention time on specific column matrix type which is being purged with a carrier gas that brings the chemical or metabolite of analytical interest to a number of possible detection systems including flame, electron capture, and mass spectrometer type (Andrade et al., 2014; Cavalheiro et al., 2014; Gao et al., 2014; Ippolito et al., 2014; Otto et al., 2015).

3.3 Mass Spectrometry

There are a variety of mass spectrometry (MS) techniques which confer great sensitivity and the possibility of multielement analyses. The general technique is based upon measuring the distance a molecule sprayed into a charged field travels as a function of its mass or mass-to-charge ratio. This technique is highly sensitive and provides highly useful data for chemical speciation studies (Sampson et al., 1994). Combined HPLC/MS systems have proven extremely useful in increasing the sensitivity and specificity of analytical measurements for chemicals such as atrazine in complex matrices such as human urine (Kuklenyik et al., 2012).

3.4 Atomic Absorption Spectroscopy

For measurement of trace elements, atomic absorption spectroscopy is a commonly used analytical technique which measures the concentration of an element of interest following its volatilization into a burner head flame or heated graphite atomizer or cold vapor, in the case of volatile elements such as mercury, by absorption of a specific spectral wavelength produced by a lamp containing that element. This technology has been in place for many years and is now often replaced by inductively coupled plasma mass spectroscopy which permits measurement of many elements volatilized in a plasma.

3.5 X-Ray Fluorescence (XRF), Electron (EDX), and Proton-Induced X-Ray Emmission Analysis (PIXEA)

These relate technologies which utilize either incident X-rays, accelerated electrons or protons to dislodge orbital electrons from the K, L, or M energy shells of atoms in a sample with generation of element-specific K, L, or M X-rays of characteristic energy or wavelength which can be measured by either an energy dispersive or wavelength dispersive detector (Fowler, 1983; Fowler et al., 1975). The technique is rapid and may provide information via the sample localization or intracellular localization of elements of interest (Figs. 2.1 and 2.2). These data

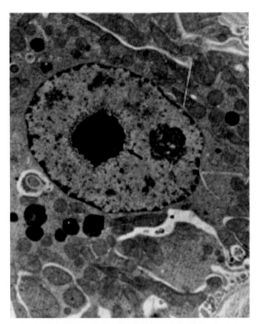

FIGURE 2.1 Electronmicrograph of a pathognomonic lead intranuclear inclusion body in a renal tubule cell from a rat exposed to a lead-containing diet. *Reproduced from Mahaffey and Fowler (1977).*

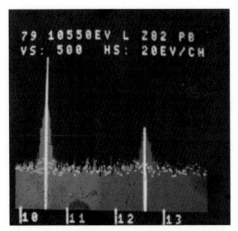

FIGURE 2.2 Energy dispersive X-ray spectrum from a lead intranuclear inclusion body showing the marked lead L alpha and L beta peaks and a superimposed spectra from an adjacent area of cytoplasm to illustrate the greater relative concentration lead concentration in the inclusion body. *Reproduced from Fowler et al. (1980).*

are highly useful in helping to interpret total elemental analyses on bulk tissue samples since they may indicate in which cells and where in affected cells the element of interest is localized.

4 BIOMONITORING STUDIES

Given the array of sophisticated and evolving analytical methodologies noted earlier, there have been a number of published biomonitoring studies employing these methods for public health purposes. Data from these studies have provided useful human exposure assessment data for a number of chemicals. These data provide critical information for conducting improved risk assessments and helping to identify subpopulations at special risk for adverse outcomes.

4.1 NHANES Studies

Among the oldest, most extensive and well-known biomonitoring studies are the National Health and Nutrition Examination Survey (NHANES). Please see Crinnion (2010). This extensive body of information has proven invaluable in documenting the public health impacts of lead removal from gasoline and the uptake of cadmium from dietary sources (Ruiz et al., 2010 and Fig. 2.3).

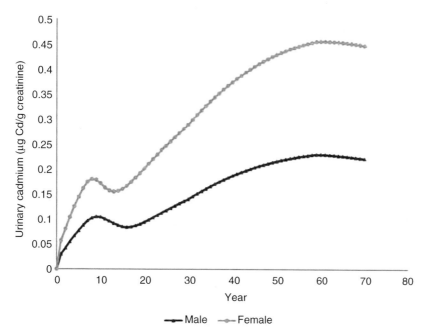

FIGURE 2.3 Urinary cadmium concentration data from the NHANES illustrating the marked differences between males and females by age and the bimodal nature of the excretion pattern as a function of age. *From Ruiz et al. (2010).*

4.2   Other National/International Biomonitoring Programs

Other national/international biomonitoring programs exist in Europe under the auspices of various national agencies. In addition, there are national biomonitoring programs in Australia (Aylward et al., 2014; Van den Eede et al., 2015), Canada (St-Amand et al., 2014), China, and 17 European countries. Regardless of the program sponsorship, all of these efforts share a strong commitment to the generation of high quality analytical biomonitoring data through in-place QA/QC programs which may include use of Standard Reference Materials from the US National Institute of Standards and Technology. Despite these large efforts, there are still many chemicals in commerce and more entering every year about which we know little. The issue of chemical mixtures and more recently nanomaterials further complicate regulatory decision making. Nonetheless, solid biomonitoring data are critical components for conducting high quality risk assessments. It is very clear that computational modeling approaches are essential tools for dealing with the large and complex data sets generated by modern analytical chemistry methods and supporting credible risk assessments for individual chemicals or complex mixtures.

5   BIOLOGICAL MONITORING FOR CHEMICAL METABOLITES AND INTERCONVERTED CHEMICAL SPECIES

In addition to accurate measurement of parent compounds, it is important to be able to measure chemical metabolites produced by in vivo metabolism. Oxidation, reduction, conjugation, and methylation reactions in vivo generate new chemical species which may have toxic properties greater or lesser than the parent compounds, thus further complicating accurate risk assessment. In addition, there is growing recognition of the importance of reactive chemical species (ROS) such as oxygen or thiyl radicals formed during metabolism which are actual toxic entities. Such information is critical for mechanism or mode of action risk assessment since the degree to which these radical may be formed may be influenced by gender, genetic inheritance, age, or nutritional factors (Fowler, 2012). The good news with regard to the complexity of these data and interactive factors is that they are well recognized as important in determining the shape of a dose-response curve for a specific subset of the population. Clearly, the utilization of computational methods will be of great value in analyzing and interpreting biomonitoring data for risk assessment purposes.

6   CLINICAL BIOMARKERS—CURRENT USAGES AND PROSPECTS FOR THE FUTURE

A noted earlier, there are a number of clinical biomarkers in current usage when a person visits a primary care provider for a routine physical examination. These biomarkers which have long history in clinical medicine tend to be enzyme activities or important metabolic substances such as glucose. As noted earlier, most of the clinical markers in current use such as LDH, SGOT, SGPT, PSA, etc. are markers of tissue injury which reflect dead and dying cells releasing these enzymes into the blood or urine and generally reflect a damaging process that is fairly well established with large number of cells being affected and extensive tissue damage.

More recently developed biomarkers that are capable of detecting ongoing toxic processes at earlier stages of development include enzymes such as δ-aminolevulinic acid dehydratase (ALAD). This enzyme which catalyzes the second step in the heme biosynthetic pathway is highly sensitive to lead inhibition and is the major carrier protein for lead in blood (Bergdahl et al., 1998; Tian et al., 2013). This Zn-dependent enzyme exists as two isomers (ALAD-1 and ALAD-2). These two isomers vary in their binding affinity for lead and have been associated with differences in susceptibility to lead-induced hypertension in the general US population (Scinicariello et al., 2010). Lead inhibition of ALAD has been found to be regulated by zinc donation from metallothionein (MT) (Goering and Fowler, 1987a,b) and zinc donation and lead chelation by low molecular weight lead-binding proteins found in major target organs such as the kidney (Goering and Fowler, 1984, 1985) and brain (Goering et al., 1986). The point of this discussion is that biomonitoring for carrier proteins has provided useful mechanism-based risk assessment information that helps identify why some subpopulations are at special risk for lead toxicity as a function of Zn nutritional status and genetic inheritance on a molecular level. The finding of lead bound to similar proteins in humans (Quintanilla-Vega et al., 1995; Smith et al., 1998) with environmental lead exposure levels suggests that these molecules may play important regulatory roles in mediating lead toxicity at low dose exposures.

Metabolomics is another approach to biological monitoring which has proven useful in epidemiological studies of persons with elevated exposures to arsenic in drinking water in Mexico. In this case arsenical inhibition of several steps in the heme biosynthetic pathway with resultant uroporphyrinuria/coproporphyrinuria (Garcia-Vargas et al., 1994) was closely associated with urinary arsenic concentrations. Similar findings have been found in patients exposed to arsenic from burning of coal in China (Xie et al., 2001). These data (Woods and Fowler, 1977b, 1978) which were in agreement with prior experimental animal studies (Woods and Fowler, 1977a, 1978) provided mechanism-based risk assessment data linking measured exposures to an adverse outcome endpoint (porphyrinuria). These metabolic effects were linked with ultrastructural morphometric and other alterations of mitochondrial respiratory function (Fowler and Woods, 1977b, 1979; Fowler et al., 1979)

Another biomonitoring study which utilized the NHANES data set (Ruiz et al., 2010) employed Madonna software to identify females in the 6–11 age range as being at special risk for increased uptake of cadmium from the diet (Fig. 2.4).

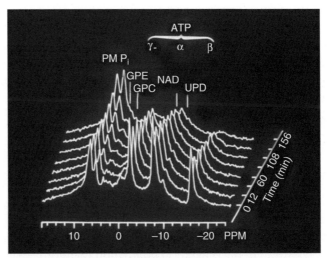

FIGURE 2.4 In vivo 31-P spectra from a rat given a single acute dose of arsenite showing the loss of ATP formation over time and corresponding increases in phophorylation of low molecular species and inorganic phosphorous. *From Chen et al. (1986).*

The value of these data rests with indicating increased uptake of cadmium at an early life stage and suggests possible nutritional interventions. Combining these biological monitoring data analyses with studies of cadmium-induced proteinuria patterns involving kidney injury molecule-1, which is a constituent protein of the renal tubular cell membrane (Prozialeck et al., 2009a,b) and is released into the urine at early stage of toxicity (Bailly et al., 2002) should provide improved mechanism-based risk assessment information to link low-dose cadmium exposures in the general population and help to identify subpopulations at special risk for long-term kidney injury over the course of a lifetime.

The major point of this chapter is to provide examples of how developing linkages between analytical biomonitoring data for specific chemicals in human populations and sensitive sentinel biological molecules or metabolic systems may lead to improved public health risk assessments and identify subpopulations at special risk for chemical-induced toxicity.

6.1 Altered Metabolite Profiles

Exposure to chemical agents or drugs or metabolic diseases such as diabetes produce alterations in normal proteins that are of potential utility as diagnostic molecular markers. One well-known example would be glycosylation of hemoglobin to produce HbAc which is used as marker of diabetes but can also be produced by increased ingestion of aspirin (Xu et al., 2000). This means that interpretation of this putative biomarker must include a consideration of other possible alkylation sources.

In a similar manner, acute in vivo exposure to arsenic (As^{3+}) has been shown by ^{31}P-NMR to produce decreased ATP formation (Fig. 2.4) but results in the increased formation of ^{31}P-phosphorlylated proteins and other ^{31}P low molecular weight species (Chen et al., 1986). These data could suggest that decreased formation of ATP by arsenic exposure increases the intracellular pool of phosphate which then has increased bioavailability for phosphorylation of other molecules such as proteins involved in important cellular regulation pathways. Further research is clearly needed in this area with regard to arsenic-induced carcinogenesis, diabetes, and apoptosis.

6.2 RIAs for Unique Proteins Resulting From Chemical or Pharmaceutical Exposures

As indicated earlier, chemical, metabolic, or pharmaceutical exposures may result in posttranslational alterations of proteins involved in a number of essential cellular functions including regulation of major metabolic pathways. Once the biological impact of these modifications is understood, it is usually more economical and efficient to monitor the formation of these altered proteins with monoclonal antibodies via their specific epitopes than to engage in mass spectroscopic analyses. The RIA for these proteins then becomes the biomarker test for predicting the biological outcome of the protein medication.

6.3 Gene Expression Patterns

It has been appreciated for a number of years that exposure of cells to chemical agents or pharmaceuticals will usually result in both up- and downregulation of a number of genes. Some of these are "housekeeping" genes which are normally operating at basal levels, others may be "stress protein genes" which are upregulated in response to chemical exposures and oncogenes that may be involved in altered regulation of the cell cycle.

6.4 Genomic/Proteomic/Metabolomic Profiling

Chemical- or pharmaceutical-induced alterations in gene expression patterns will have obvious implications for other related downstream biological processes such as the proteome and metabolome. All these important general sets of processes are interconnected and feed back to each other. Ideally one would like to have an overall appreciation of how these interrelationships interact with each other as a function of dose, time, age, and gender. Under-control and chemical/pharmaceutical exposures has given rise to the growing field of systems biology which is highly complex but provides useful insights into how these systems interact with each other under a variety of normal and stress situations. The evolution of computerized systems biology approaches has provided valuable data on possible interactions between these biological systems which can lead to hypotheses that may be tested in under-laboratory conditions. A further interesting new area of research concerns the role of the microbiome/microflora (Schneider and Winslow, 2014) of the gastrointestinal tract in mediating metabolic transformations in vivo. The impact of these microbes on biomarker systems and measured endpoints is an open area of current investigation.

7 BIOMARKER MODIFYING FACTORS AND IDENTIFICATION OF POPULATIONS AT RISK

As noted earlier, there are a number of general factors which may alter the response of a given putative biomarker in either a qualitative or quantitative manner and hence influence interpretation of the biomarker for risk assessment purposes. Among these factors are genetic, epigenetic, age, nutritional, inducible metabolic systems, and chemical/drug mixture impacts on measured biomarker responses. These interactive influences may greatly alter the shape of an expected biomarker dose-response curve and ultimate health outcome. These factors can define, to large degree, populations at special risk for toxicity by addressing the issue of "dose of a chemical or pharmaceutical agent to whom."

7.1 Genetic Factors

Genetic factors clearly help to define us as individuals and our intrinsic susceptibility to toxicity from chemicals or pharmaceutical agents (Chang et al., 2009). The impact of genetic factors in this area will be further modified by gender, age (life stage), and nutritional status. So ideally all these aspects should be considered together in order to achieve the most robust assessment of risk from an exposure or group of exposures. A good example of this would be the finding of increased susceptibility to lead-induced hypertension among carriers of the ALAD-2 allele (Scinicariello et al., 2010). This kind of genetic information is extremely valuable to define the subpopulation at special risk for a major adverse health effect induced by environmental exposure to a ubiquitous toxic agent.

7.2 Epigenetic Factors

Epigenetic factors such as altered regulation of genes in central metabolic or adverse outcome pathways (AOPs) by gene methylation or micro RNAs have been recognized in recent years (Gadola et al., 2014) as playing important roles in mediating the development of cell injury/cell death or cancer. Data from these studies have clearly shown gene activation or silencing play major roles in determining sensitivity to chemicals or pharmaceuticals and ultimately health outcomes. This is an active area of current research which will be discussed in later chapters of this book.

7.3 Inducible Metabolic/Toxicant Regulation Systems

Cells have a number of inducible systems which may act to protect or facilitate cell injury processes. Among the better studied are the cytochrome-450s, Phase II conjugation enzymes, the MT family of proteins, and antioxidant systems such as glutathione. The importance of these inducible systems rests with their ability to not only help define thresholds of cellular toxicity but also the shape of a dose-response curve and whether it is linear or multiphasic. These considerations have clear importance in risk assessment decision making.

7.3.1 Cytochrome P-450 Family of Enzymes and Phase II Enzymes

Cytochrome P-450 family of hemoproteins has been studied for many years and is recognized to play a central role in the metabolism of both endogenous and xenobiotic organic chemicals. These metabolic conversions may result in both toxication reactions from generation of reactive oxygen species (ROS) and detoxication reactions which render the chemical of interest more hydrophilic and hence is more readily excreted from the body following conjugation with glutathione or glucuronic acid or other small molecules (Groer et al., 2014; Li et al., 2014; Reinen and Vermeulen, 2015; Zang et al., 2014).

7.3.2 Metallothionein Family of Metal-Binding Proteins

MT is a protein which has a large gene family whose members are found across a number of phyla (Fowler et al., 1987a) and code for low molecular weight cysteine-rich proteins which may be induced by a number of metals (Andrews, 2000; Miura et al., 1998) and oxidative stress (Liu et al., 2004; Provinciali et al., 2002; Shibuya et al., 2008). There are also a number of isoforms of MT and MT has been shown to be able to donate essential metals such as zinc to zinc-dependent enzymes such as ALAD and to attenuate the inhibitory effects of lead on this biomarker for lead. This has obvious implications for risk assessments of combined exposures for lead with other metals/metallics (Andrade et al., 2014; Madden et al., 2002; Madden and Fowler, 2000; Mahaffey et al., 1981; Rodriguez-Sastre et al., 2014; Whittaker et al., 2011). This issue will be taken up later in sections dealing with mixture exposures (Please see Chapter 5, section 2.4). The proteins play both protective roles on an intracellular basis in binding toxic metals such cadmium and as a defense against ROS (Shibuya et al., 2008). On the other hand, these molecules may facilitate adverse health outcomes by transporting toxic metals between organ systems and attenuating the removal of metals such as cadmium from the body resulting in a prolonged biological half-life. Attempts to utilize MT as a molecular biomarker for prediction of susceptibility to toxicity have met with limited success since the roles played by this molecule in mediating AOPs are relatively unknown.

7.3.3 Antioxidant Systems

Many toxic agents including both metals (Mohammadi-Bardbori and Rannug, 2014) and organic chemicals (Zhang et al., 2014) induce cell injury/cell death via generation of ROS (Engel and Evens, 2006) as a result of action on the mitochondria (Zorov et al., 2014) and endoplasmic reticulum-based enzyme systems (Goswami et al., 2014; Jin et al., 2014; Yang et al., 2014b; Zhang et al., 2014, 2015). Intracellular ROS in large quantities may damage cellular organelles and disrupt the function of essential biochemical systems. Fortunately, there are a number of inducible intracellular antioxidant systems which help to regulate ROS concentrations and maintain homeostasis. Glutathione, which is present in most cells at millimolar concentrations, is a major antioxidant regulatory element. Intracellular concentrations of this SH-containing molecule may be increased by induction of the enzyme glutathione synthetase. As noted earlier, the SH-rich protein MT may also act as an antioxidant and scavenge ROS. The main point here is that cells possess intracellular protective systems whose presence will impact both the measurement and interpretation of putative biomarkers responding to ROS until their capacity is exceeded. This is an important concept with regard to the utilization of molecular biomarkers for risk assessment since it appears overt chemical-induced cell injury and ultimately clinical organ disease only usually occur once the capacity of these cellular defense systems to provide protection is exceeded. In this sense, these protective cellular systems play a major role in defining populations at special risk for toxicity. As with the molecular biomarker endpoints the capacity of antioxidant systems is also influenced by factors such as genetic inheritance, age (Fowler and Woods, 1977a), gender (Fowler, 2005; Fowler et al., 2008), and nutritional status (Ojo et al., 2014). Hence, in order to use molecular biomarkers for risk assessment purposes, all these interactive factors should be included in any overall evaluation.

8 TECHNICAL ADVANCES IN INSTRUMENTATION

8.1 Automated and Robotic Analytical Systems

Clearly the increasing array of possible molecular biomarker systems and growing array of possible interactive elements require sophisticated analytical methodologies which will generate large quantities of data which must be evaluated. Evolving analytical methodologies increasingly rely on robotic handling of samples and on-board computer management systems to both manage the analyses and process the generated data. This is particularly important for high-throughput systems such as those utilized in NHANES analytical data sets (Aylward et al., 2013), Tox21 (Attene-Ramos et al., 2014; Judson et al., 2013; Shaughnessy et al., 2014), and ToxCast (Filer et al., 2014; Wetmore et al., 2014) approaches for biomonitoring and biomarker development. It needs to be further emphasized that computational analysis is essential in order to digest the large quantities of data generated by these technologies and provide the scientific foundation for systems biology/toxicology-based risk assessments which is the ultimate goal of research in this area to improve risk assessments for toxic chemicals or pharmaceutical agents.

8.2 Mircrofluidics

One of the increasingly important advancements in the field of analytical chemistry has been the development of technical systems capable of accurately analyzing increasingly small biological samples. These methodologies have permitted analysis of tissue samples collected by small tissue sample biopsies and other small sample volumes (Gallagher et al., 2014; Jain, 2007, 2008, 2012) and increased the number of analyses that may be conducted on limited sample volumes such as those collected for the NHANES survey (Sampson et al., 1994). These advances in analytical technology have permitted the development of the field of nanoproteomics which will be discussed in greater detail in subsequent chapters.

8.3 Computer Management Systems

The previously mentioned advances in analytical methodologies require computer management systems to handle sample processing prior to analysis, conduct measurements via operation of sophisticated robotic analytical devices, and process the large quantities of data generated. Clearly these essential tasks have been greatly aided by the evolution of microprocessor-based computer technologies which have lead to the development of on-board computers in modern analytical devices such as HPLC/mass spectrometers in combination (Hartler et al., 2007; Ubaida Mohien et al., 2010). The evolution of artificial intelligence systems can only be expected to accelerate the production of large quantities of analytical data in future.

9 BASIC SCIENTIFIC BIOMARKER VALIDATION APPROACHES

A central question which arises from the generation of large putative biomarker analytical data sets is how can this information be utilized to produce useful human health risk assessments. In order for this translation to occur, the analytical data must be validated not only in terms of analytical chemistry with appropriate QA/QC but also correlated with other biological parameters of toxicity so that a correct prognostic interpretation may be placed upon the findings. In other words, what do these findings mean in terms of probable future health outcomes? This is not a simple task and requires integration of another large set of correlative parameters at both the basic cellular and intact organism levels of biological organization. Ultimately, these studies would support a systems biology/toxicology-based approach to risk assessment for individual chemicals or chemical mixtures. Computational methodologies appear to be the only viable approach for integrating these large and diverse data sets (Tox21, ToxCast)

9.1 Correlations With Histopathology and Electron Microscopy

At the cellular level, correlation of putative biomarkers with morphological manifestations of toxicity is an important early step in the biomarker validation process. Alterations in cellular morphology have been used for many years to predict health outcomes and form the basis of histopathological evaluations of cancers and other adverse health outcomes. At the ultrastructural level, alterations in the structure of organelles such as the mitochondria (Fowler, 1983), endoplasmic reticulum (Fowler et al., 1977, 1983), lysosomes (Fowler et al., 1975), and peroxisomes (Kondo and Makita, 1997) have provided important insights into likely underlying causes of cellular toxicity from chemical exposures. Application of ultrastructural morphometric approaches (Fowler, 1983) has permitted quantification of chemical-induced alterations of these subcellular systems (eg, mitochondria) which has been correlated with changes in putative biomarkers such as specific porphyrinuria patterns (Fowler et al., 1979). Changes in porphyrin excretion patterns can be interpreted in a quantitative manner at the level of the target cell population. This combined ultrastructural morphometric/biochemical approach has proven extremely valuable in terms of providing basic scientific information in support of mode of action risk assessments for agents such as arsenic (Fowler et al., 1987b, 1979), methyl mercury (Fowler et al., 1975; Fowler and Woods, 1977a), and cadmium (Squibb et al., 1984) in liver and kidney. There will be a more detailed discussion of the roles of this cell biology approach to toxicology in providing a mechanistic foundation for placing systems biology information and attendant risk assessment applications in a target cell/intracellular context. Ultimately the net result of linking these interrelated data sets will be a more comprehensive appreciation of the relationships which must exist between toxicant-induced disturbances in AOPs and alterations in the structure of organelles housing those biochemical pathways. In other words, combining these lines of information will ultimately lead to a more comprehensive and interpretable picture of the ongoing toxic process. Synthesis of these complex data sets will obviously require computer-assisted analyses.

9.2 In Vivo/In Vitro Cell Biology Comparison Studies

In recent years, increasing numbers of toxicology studies have utilized in vitro test systems to generate basic insights into mechanisms of toxicant action at the cellular level of biological organization (Madden and Fowler, 2000). The Tox21 and ToxCast initiatives have utilized high-throughput approaches to develop in vitro toxicology information on a large number of chemical agents. In order for these data to be of value for risk assessment purposes, they must be compared and validated against findings from in vivo studies and further translated into the human context by computational modeling approaches. This is a complex and labor-intensive undertaking but given the progress made in recent years in these areas, it appears that it may be feasible in the next decade. A conceptual model for how this translational process which ranges from the molecular to the cellular level, intact animal levels of biological organization, followed by computational modeling translation could be implemented for human health risk assessments.

This approach would ideally provide a scientifically defensible foundation for systems biology–based risk assessments in the future by examining cellular responses to toxic agent exposures across the various levels of biological organization ranging from the molecular to organelle to target cell to intact experimental animal to exposed human populations. It is important to note that with the increasing availability of human genetic data from NHANES (Chang et al., 2009) it should be increasingly possible to factor genetic inheritance into the risk assessment process for both AOPs and epigenetic polymorphisms which may influence overall health outcomes. The net result would ideally be generation of personalized health risk assessments and hence increased protection for sensitive subpopulations at special risk for toxicity from both specific chemicals and mixtures of chemicals.

Based upon the earlier discussion, personalized human health risk assessments would seem to be a worthy and increasingly feasible future goal which would take advantage of the diverse range of modern tools being applied to understand mechanisms of chemical-induced toxicity and translated this basic scientific information into risk assessment practice. Subsequent chapters of this book will delve, in greater detail, into more specific aspects of these important interrelated components.

References

Amayo, K.O., Raab, A., Krupp, E.M., Feldmann, J., 2014. Identification of arsenolipids and their degradation products in cod-liver oil. Talanta 118, 217–223.

Andrade, V., Mateus, M.L., Batoreu, M.C., Aschner, M., Marreilha Dos Santos, A.P., 2014. Changes in rat urinary porphyrin profiles predict the magnitude of the neurotoxic effects induced by a mixture of lead, arsenic and manganese. Neurotoxicology 45, 168–177.

Andrews, G.K., 2000. Regulation of metallothionein gene expression by oxidative stress and metal ions. Biochem. Pharmacol. 59 (1), 95–104.

Attene-Ramos, M.S., Huang, R., Michael, S., Witt, K.L., Richard, A., Tice, R.R., et al., 2014. Profiling of the Tox21 chemical collection for mitochondrial function to identify compounds that acutely decrease mitochondrial membrane potential. Environ. Health Perspect. 123, 49–56.

Aylward, L.L., Kirman, C.R., Schoeny, R., Portier, C.J., Hays, S.M., 2013. Evaluation of biomonitoring data from the CDC National Exposure Report in a risk assessment context: perspectives across chemicals. Environ. Health Perspect. 121 (3), 287–294.

Aylward, L.L., Green, E., Porta, M., Toms, L.M., Den Hond, E., Schulz, C., et al., 2014. Population variation in biomonitoring data for persistent organic pollutants (POPs): an examination of multiple population-based datasets for application to Australian pooled biomonitoring data. Environ. Int. 68, 127–138.

Bailly, V., Zhang, Z., Meier, W., Cate, R., Sanicola, M., Bonventre, J.V., 2002. Shedding of kidney injury molecule-1, a putative adhesion protein involved in renal regeneration. J. Biol. Chem. 277 (42), 39739–39748.

Bergdahl, I.A., Sheveleva, M., Schutz, A., Artamonova, V.G., Skerfving, S., 1998. Plasma and blood lead in humans: capacity-limited binding to delta-aminolevulinic acid dehydratase and other lead-binding components. Toxicol. Sci. 46 (2), 247–253.

Cavalheiro, J., Preud'homme, H., Amouroux, D., Tessier, E., Monperrus, M., 2014. Comparison between GC-MS and GC-ICPMS using isotope dilution for the simultaneous monitoring of inorganic and methyl mercury, butyl and phenyl tin compounds in biological tissues. Anal. Bioanal. Chem. 406 (4), 1253–1258.

Chang, M.H., Lindegren, M.L., Butler, M.A., Chanock, S.J., Dowling, N.F., Gallagher, M., et al., 2009. Prevalence in the United States of selected candidate gene variants: Third National Health and Nutrition Examination Survey, 1991–1994. Am. J. Epidemiol. 169 (1), 54–66.

Chen, B., Burt, C.T., Goering, P.L., Fowler, B.A., London, R.E., 1986. In vivo 31P nuclear magnetic resonance studies of arsenite induced changes in hepatic phosphate levels. Biochem. Biophys. Res. Commun. 139 (1), 228–234.

Crinnion, W.J., 2010. The CDC fourth national report on human exposure to environmental chemicals: what it tells us about our toxic burden and how it assist environmental medicine physicians. Altern. Med. Rev. 15 (2), 101–109.

Engel, R.H., Evens, A.M., 2006. Oxidative stress and apoptosis: a new treatment paradigm in cancer. Front. Biosci. 11, 300–312.

Filer, D., Patisaul, H.B., Schug, T., Reif, D., Thayer, K., 2014. Test driving ToxCast: endocrine profiling for 1858 chemicals included in phase II. Curr. Opin. Pharmacol. 19, 145–152.

Fowler, B.A., 1983. Role of ultrastructural techniques in understanding mechanisms of metal-induced nephrotoxicity. Fed. Proc. 42 (13), 2957–2964.

Fowler, B.A., 2005. Molecular biomarkers: challenges and prospects for the future. Toxicol. Appl. Pharmacol. 206 (2), 97.

Fowler, B.A., 2012. Biomarkers in toxicology and risk assessment. EXS 101, 459–470.

Fowler, B.A., Woods, J.S., 1977a. The transplacental toxicity of methyl mercury to fetal rat liver mitochondria. Morphometric and biochemical studies. Lab. Invest. 36 (2), 122–130.

Fowler, B.A., Woods, J.S., 1977b. Ultrastructural and biochemical changes in renal mitochondria during chronic oral methyl mercury exposure: the relationship to renal function. Exp. Mol. Pathol. 27 (3), 403–412.

Fowler, B.A., Woods, J.S., 1979. The effects of prolonged oral arsenate exposure on liver mitochondria of mice: morphometric and biochemical studies. Toxicol. Appl. Pharmacol. 50 (2), 177–187.

Fowler, B.A., Brown, H.W., Lucier, G.W., Krigman, M.R., 1975. The effects of chronic oral methyl mercury exposure on the lysosome system of rat kidney. Morphometric and biochemical studies. Lab. Invest. 32 (3), 313–322.

Fowler, B.A., Hook, G.E., Lucier, G.W., 1977. Tetrachlorodibenzo-*p*-dioxin induction of renal microsomal enzyme systems: ultrastructural effects on pars recta (S3) proximal tubule cells on the rat kidney. J. Pharmacol. Exp. Ther. 203 (3), 712–721.

Fowler, B.A., Woods, J.S., Schiller, C.M., 1979. Studies of hepatic mitochondrial structure and function: morphometric and biochemical evaluation of in vivo perturbation by arsenate. Lab. Invest. 41 (4), 313–320.

Fowler, B.A., Kimmel, C.A., Woods, J.S., McConnell, E.E., Grant, L.D., 1980. Chronic low level lead toxicity in the rat III An integrated toxicological assessment with special reference to the kidney. Toxicol. Appl. Pharmacol. 56, 59–77.

Fowler, B.A., Kardish, R.M., Woods, J.S., 1983. Alteration of hepatic microsomal structure and function by indium chloride. Ultrastructural, morphometric, and biochemical studies. Lab. Invest. 48 (4), 471–478.

Fowler, B.A., Hildebrand, C.E., Kojima, Y., Webb, M., 1987a. Nomenclature of metallothionein. Experientia Suppl. 52, 19–22.

Fowler, B.A., Oskarsson, A., Woods, J.S., 1987b. Metal- and metalloid-induced porphyrinurias. Relationships to cell injury. Ann. NY Acad. Sci. 514, 172–182.

Fowler, B.A., Conner, E.A., Yamauchi, H., 2008. Proteomic and metabolomic biomarkers for III-V semiconductors: and prospects for application to nano-materials. Toxicol. Appl. Pharmacol. 233 (1), 110–115.

Gadola, L., Noboa, O., Staessen, J.A., Boggia, J., Striteska, J., Nekvindova, J., et al., 2014. MicroRNAs and kidneys. Int. J. Nephrol. 153 (4), 187–192.

Gallagher, R., Dillon, L., Grimsley, A., Murphy, J., Samuelsson, K., Douce, D., 2014. The application of a new microfluidic device for the simultaneous identification and quantitation of midazolam metabolites obtained from a single micro-litre of chimeric mice blood. Rapid Commun. Mass Spectrom. 28 (11), 1293–1302.

Gambrill, E., Shlonsky, A., 2000. Risk assessment in context. Children Youth Serv. Rev. 22 (11–12), 813–837.

Gao, J., Zhang, K., Chen, Y., Guo, J., Wei, Y., Jiang, W., et al., 2014. Occurrence of organotin compounds in river sediments under the dynamic water level conditions in the Three Gorges Reservoir Area. China. Environ. Sci. Pollut. Res. Int. 22, 8375–8385.

Garcia-Vargas, G.G., Del Razo, L.M., Cebrian, M.E., Albores, A., Ostrosky-Wegman, P., Montero, R., et al., 1994. Altered urinary porphyrin excretion in a human population chronically exposed to arsenic in Mexico. Hum. Exp. Toxicol. 13 (12), 839–847.

Goering, P.L., Fowler, B.A., 1984. Regulation of lead inhibition of delta-aminolevulinic acid dehydratase by a low molecular weight, high affinity renal lead-binding protein. J. Pharmacol. Exp. Ther. 231 (1), 66–71.

Goering, P.L., Fowler, B.A., 1985. Mechanism of renal lead-binding protein reversal of delta-aminolevulinic acid dehydratase inhibition by lead. J. Pharmacol. Exp. Ther. 234 (2), 365–371.

Goering, P.L., Fowler, B.A., 1987a. Kidney zinc-thionein regulation of delta-aminolevulinic acid dehydratase inhibition by lead. Arch. Biochem. Biophys. 253 (1), 48–55.

Goering, P.L., Fowler, B.A., 1987b. Metal constitution of metallothionein influences inhibition of delta-aminolaevulinic acid dehydratase (porphobilinogen synthase) by lead. Biochem. J. 245 (2), 339–345.

Goering, P.L., Mistry, P., Fowler, B.A., 1986. A low molecular weight lead-binding protein in brain attenuates lead inhibition of delta-aminolevulinic acid dehydratase: comparison with a renal lead-binding protein. J. Pharmacol. Exp. Ther. 237 (1), 220–225.

Goswami, P., Gupta, S., Biswas, J., Joshi, N., Swarnkar, S., Nath, C., Singh, S., 2014. Endoplasmic reticulum stress plays a key role in rotenone-induced apoptotic death of neurons. Mol. Neurobiol. 53, 285–298.

Groer, C., Busch, D., Patrzyk, M., Beyer, K., Busemann, A., Heidecke, C.D., et al., 2014. Absolute protein quantification of clinically relevant cytochrome P450 enzymes and UDP-glucuronosyltransferases by mass spectrometry-based targeted proteomics. J. Pharm. Biomed. Anal. 100, 393–401.

Hartler, J., Thallinger, G.G., Stocker, G., Sturn, A., Burkard, T.R., Korner, E., et al., 2007. MASPEC-TRAS: a platform for management and analysis of proteomics LC-MS/MS data. BMC Bioinformatics, 8, 197.

Ippolito, D.L., Lewis, J.A., Yu, C., Leon, L.R., Stallings, J.D., 2014. Alteration in circulating metabolites during and after heat stress in the conscious rat: potential biomarkers of exposure and organ-specific injury. BMC Physiol. 14 (1), 1.

Jain, K.K., 2007. Cancer biomarkers: current issues and future directions. Curr. Opin. Mol. Ther. 9 (6), 563–571.

Jain, K.K., 2008. Nanomedicine: application of nanobiotechnology in medical practice. Med. Princ. Pract. 17 (2), 89–101.

Jain, K.K., 2012. Role of nanodiagnostics in personalized cancer therapy. Clin. Lab. Med. 32 (1), 15–31.

Jain, R.B., Wang, R.Y., 2011. Association of caffeine consumption and smoking status with the serum concentrations of polychlorinated biphenyls, dioxins, and furans in the general U.S. population: NHANES 2003–2004. J. Toxicol. Environ. Health A 74 (18), 1225–1239.

Jin, Y., Zhang, S., Tao, R., Huang, J., He, X., Qu, L., Fu, Z., 2014. Oral exposure of mice to cadmium (II), chromium (VI) and their mixture induce oxidative- and endoplasmic reticulum-stress mediated apoptosis in the livers. Environ. Toxicol. Nov, 22082.

Judson, R., Kavlock, R., Martin, M., Reif, D., Houck, K., Knudsen, T., et al., 2013. Perspectives on validation of high-throughput assays supporting 21st century toxicity testing. ALTEX 30 (1), 51–56.

Kondo, K., Makita, T., 1997. Morphometry of abnormal peroxisomes induced by withdrawal of bezafibrate, a hypolipidemic drug in male rat hepatocytes. J. Vet. Med. Sci. 59 (4), 297–299.

Krewski, D., Westphal, M., Andersen, M.E., Paoli, G.M., Chiu, W.A., Al-Zoughool, M., et al., 2014. A framework for the next generation of risk science. Environ. Health Perspect. 122 (8), 796–805.

Kuklenyik, Z., Panuwet, P., Jayatilaka, N.K., Pirkle, J.L., Calafat, A.M., 2012. Two-dimensional high performance liquid chromatography separation and tandem mass spectrometry detection of atrazine and its metabolic and hydrolysis products in urine. J. Chromatogr. B Analyt. Technol. Biomed. Life Sci. 901, 1–8.

Li, S., Dong, J.Y., Guo, C.G., Wu, Y.X., Zhang, W., Fan, L.Y., et al., 2013. A stable and high-resolution isoelectric focusing capillary array device for micropreparative separation of proteins. Talanta 116, 259–265.

Li, J., Zhang, D., Jefferson, P.A., Ward, K.M., Ayene, I.S., 2014. A bioactive probe for glutathione-dependent antioxidant capacity in breast cancer patients: implications in measuring biological effects of arsenic compounds. J. Pharmacol. Toxicol. Methods 69 (1), 39–48.

Liu, A.L., Zhang, Z.M., Zhu, B.F., Liao, Z.H., Liu, Z., 2004. Metallothionein protects bone marrow stromal cells against hydrogen peroxide-induced inhibition of osteoblastic differentiation. Cell Biol. Int. 28 (12), 905–911.

Luong, J., Gras, R., Hawryluk, M., Shellie, R.A., Cortes, H.J., 2013. Multidimensional gas chromatography using microfluidic switching and low thermal mass gas chromatography for the characterization of targeted volatile organic compounds. J. Chromatogr. A 1288, 105–110.

Madden, E.F., Fowler, B.A., 2000. Mechanisms of nephrotoxicity from metal combinations: a review. Drug Chem. Toxicol. 23 (1), 1–12.

Madden, E.F., Akkerman, M., Fowler, B.A., 2002. A comparison of 60, 70, and 90 kDa stress protein expression in normal rat NRK-52 and human HK-2 kidney cell lines following in vitro exposure to arsenite and cadmium alone or in combination. J. Biochem. Mol. Toxicol. 16 (1), 24–32.

Mahaffey, K.R., Fowler, B.A., 1977. Effects of concurrent administration of diet of lead, cadmium, and arsenic in the rat. Environ. Health Perspect. 19, 165–171.

Mahaffey, K.R., Capar, S.G., Gladen, B.C., Fowler, B.A., 1981. Concurrent exposure to lead, cadmium, and arsenic. Effects on toxicity and tissue metal concentrations in the rat. J. Lab. Clin. Med. 98 (4), 463–481.

Miller, T.H., McEneff, G.L., Brown, R.J., Owen, S.F., Bury, N.R., Barron, L.P., 2014. Pharmaceuticals in the freshwater invertebrate, Gammarus pulex, determined using pulverised liquid extraction, solid phase extraction and liquid chromatography-tandem mass spectrometry. Sci. Total Environ. 511, 153–160.

Miura, N., Satoh, M., Imura, N., Naganuma, A., 1998. Protective effect of bismuth nitrate against injury to the bone marrow by gamma-irradiation in mice: possible involvement of induction of metallothionein synthesis. J. Pharmacol. Exp. Ther. 286 (3), 1427–1430.

Mohammadi-Bardbori, A., Rannug, A., 2014. Arsenic, cadmium, mercury and nickel stimulate cell growth via NADPH oxidase activation. Chem. Biol. Interact. 224c, 183–188.

Naldi, M., Baldassarre, M., Nati, M., Laggetta, M., Giannone, F.A., Domenicali, M., et al., 2014. Mass spectrometric characterization of human serum albumin dimer: a new potential biomarker in chronic liver diseases. J. Pharm. Biomed. Anal. 112, 169–175.

Nie, J., Kennedy, R.T., 2013. Capillary liquid chromatography fraction collection and postcolumn reaction using segmented flow microfluidics. J. Sep. Sci. 36 (21–22), 3471–3477.

Ojo, J.O., Oketayo, O.O., Adesanmi, C.A., Horvat, M., Mazej, D., Tratnik, J., 2014. Influence of nutritional status on some toxic and essential elements in the blood of women exposed to vehicular pollution in Ile-Ife, Nigeria. Environ. Sci. Pollut. Res. Int. 21 (2), 1124–1132.

Otto, S., Streibel, T., Erdmann, S., Sklorz, M., Schulz-Bull, D., Zimmermann, R., 2015. Application of pyrolysis-mass spectrometry and pyrolysis-gas chromatography-mass spectrometry with electron-ionization or resonance-enhanced-multi-photon ionization for characterization of crude oils. Anal. Chim. Acta 855, 60–69.

Patterson, Jr., D.G., Wong, L.Y., Turner, W.E., Caudill, S.P., Dipietro, E.S., McClure, P.C., et al., 2009. Levels in the U.S. population of those persistent organic pollutants (2003-2004) included in the Stockholm Convention or in other long range transboundary air pollution agreements. Environ. Sci. Technol. 43 (4), 1211–1218.

Pedrali, A., Tengattini, S., Marrubini, G., Bavaro, T., Hemstrom, P., Massolini, G., et al., 2014. Characterization of intact neo-glycoproteins by hydrophilic interaction liquid chromatography. Molecules 19 (7), 9070–9088.

Provinciali, M., Donnini, A., Argentati, K., Di Stasio, G., Bartozzi, B., Bernardini, G., 2002. Reactive oxygen species modulate Zn(2+)-induced apoptosis in cancer cells. Free Radic. Biol. Med. 32 (5), 431–445.

Prozialeck, W.C., Edwards, J.R., Lamar, P.C., Liu, J., Vaidya, V.S., Bonventre, J.V., 2009a. Expression of kidney injury molecule-1 (Kim-1) in relation to necrosis and apoptosis during the early stages of Cd-induced proximal tubule injury. Toxicol. Appl. Pharmacol. 238 (3), 306–314.

Prozialeck, W.C., Edwards, J.R., Vaidya, V.S., Bonventre, J.V., 2009b. Preclinical evaluation of novel urinary biomarkers of cadmium nephrotoxicity. Toxicol. Appl. Pharmacol. 238 (3), 301–305.

Qiao, L., Wang, S., Li, H., Shan, Y., Dou, A., Shi, X., Xu, G., 2014. A novel surface-confined glucaminium-based ionic liquid stationary phase for hydrophilic interaction/anion-exchange mixed-mode chromatography. J. Chromatogr. A 1360, 240–247.

Quintanilla-Vega, B., Smith, D.R., Kahng, M.W., Hernandez, J.M., Albores, A., Fowler, B.A., 1995. Lead-binding proteins in brain tissue of environmentally lead-exposed humans. Chem. Biol. Interact. 98 (3), 193–209.

Rappaport, S.M., 2011. Implications of the exposome for exposure science. J. Expos. Sci. Environ. Epidemiol. 21 (1), 5–9.

Reinen, J., Vermeulen, N.P., 2015. Biotransformation of endocrine disrupting compounds by selected phase I and phase II enzymes—formation of estrogenic and chemically reactive metabolites by cytochromes P450 and sulfotransferases. Curr. Med. Chem. 22 (4), 500–527.

Rodriguez-Sastre, M.A., Rojas, E., Valverde, M., 2014. Assessing the impact of As-Cd-Pb metal mixture on cell transformation by two-stage Balb/c 3T3 cell assay. Mutagenesis 29 (4), 251–257.

Ruiz, P., Mumtaz, M., Osterloh, J., Fisher, J., Fowler, B.A., 2010. Interpreting NHANES biomonitoring data, cadmium. Toxicol. Lett. 198 (1), 44–48.

Sampson, E.J., Needham, L.L., Pirkle, J.L., Hannon, W.H., Miller, D.T., Patterson, D.G., et al., 1994. Technical and scientific developments in exposure marker methodology. Clin. Chem. 40 (7 Pt. 2), 1376–1384.

Schneider, G.W., Winslow, R., 2014. Parts and wholes: the human microbiome, ecological ontology, and the challenges of community. Perspect. Biol. Med. 57 (2), 208–223.

Scinicariello, F., Yesupriya, A., Chang, M.H., Fowler, B.A., 2010. Modification by ALAD of the association between blood lead and blood pressure in the U.S. population: results from the Third National Health and Nutrition Examination Survey. Environ. Health Perspect. 118 (2), 259–264.

Shaughnessy, D.T., McAllister, K., Worth, L., Haugen, A.C., Meyer, J.N., Domann, F.E., et al., 2014. Mitochondria, energetics, epigenetics, and cellular responses to stress. Environ. Health Perspect. 122 (12), 1271–1278.

Shi, X., Shang, W., Wang, S., Xue, N., Hao, Y., Wang, Y., et al., 2014. Simultaneous quantification of naproxcinod and its active metabolite naproxen in rat plasma using LC-MS/MS: application to a pharmacokinetic study. J. Chromatogr. B Analyt. Technol. Biomed. Life Sci. 978–979, 157–162.

Shibuya, K., Suzuki, J.S., Kito, H., Naganuma, A., Tohyama, C., Satoh, M., 2008. Protective role of metallothionein in bone marrow injury caused by X-irradiation. J. Toxicol. Sci. 33 (4), 479–484.

Smith, D.R., Kahng, M.W., Quintanilla-Vega, B., Fowler, B.A., 1998. High-affinity renal lead-binding proteins in environmentally-exposed humans. Chem. Biol. Interact. 115 (1), 39–52.

Song, H., Adams, E., Desmet, G., Cabooter, D., 2014. Evaluation and comparison of the kinetic performance of ultra-high performance liquid chromatography and high-performance liquid chromatography columns in hydrophilic interaction and reversed-phase liquid chromatography conditions. J. Chromatogr. A 1369, 83–91.

Squibb, K.S., Pritchard, J.B., Fowler, B.A., 1984. Cadmium-metallothionein nephropathy: relationships between ultrastructural/biochemical alterations and intracellular cadmium binding. J. Pharmacol. Exp. Ther. 229 (1), 311–321.

St-Amand, A., Werry, K., Aylward, L.L., Hays, S.M., Nong, A., 2014. Screening of population level biomonitoring data from the Canadian Health Measures Survey in a risk-based context. Toxicol. Lett. 231 (2), 126–134.

Thurmann, S., Belder, D., 2014. Phase-optimized chip-based liquid chromatography. Anal. Bioanal. Chem. 406 (26), 6599–6606.

Tian, L., Zheng, G., Sommar, J.N., Liang, Y., Lundh, T., Broberg, K., et al., 2013. Lead concentration in plasma as a biomarker of exposure and risk, and modification of toxicity by delta-aminolevulinic acid dehydratase gene polymorphism. Toxicol. Lett. 221 (2), 102–109.

Ubaida Mohien, C., Hartler, J., Breitwieser, F., Rix, U., Remsing Rix, L., Winter, G.E., et al., 2010. MASPECTRAS 2: an integration and analysis platform for proteomic data. Proteomics 10 (14), 2719–2722.

Ukena, T., Matsumoto, E., Nishimura, T., Harn, J.C., Lee, C.A., Rojanapantip, L., et al., 2014. Speciation and determination of inorganic arsenic in rice using liquid chromatography-inductively coupled plasma/mass spectrometry: collaborative study. J. AOAC Int. 97 (3), 946–955.

Van den Eede, N., Heffernan, A.L., Aylward, L.L., Hobson, P., Neels, H., Mueller, J.F., Covaci, A., 2015. Age as a determinant of phosphate flame retardant exposure of the Australian population and identification of novel urinary PFR metabolites. Environ. Int. 74, 1–8.

Wang, C., Feng, R., Li, Y., Zhang, Y., Kang, Z., Zhang, W., Sun, D.J., 2014. The metabolomic profiling of serum in rats exposed to arsenic using UPLC/Q-TOF MS. Toxicol. Lett. 229 (3), 474–481.

Wetmore, B.A., Allen, B., Clewell, III, H.J., Parker, T., Wambaugh, J.F., Almond, L.M., et al., 2014. Incorporating population variability and susceptible subpopulations into dosimetry for high-throughput toxicity testing. Toxicol. Sci. 142 (1), 210–224.

Whittaker, M.H., Wang, G., Chen, X.Q., Lipsky, M., Smith, D., Gwiazda, R., Fowler, B.A., 2011. Exposure to Pb, Cd, and As mixtures potentiates the production of oxidative stress precursors: 30-day, 90-day, and 180-day drinking water studies in rats. Toxicol. Appl. Pharmacol. 254 (2), 154–166.

Wild, C.P., Scalbert, A., Herceg, Z., 2013. Measuring the exposome: a powerful basis for evaluating environmental exposures and cancer risk. Environ. Mol. Mutagen. 54 (7), 480–499.

Woods, J.S., Fowler, B.A., 1977a. Effects of chronic arsenic exposure on hematopoietic function in adult mammalian liver. Environ. Health Perspect. 19, 209–213.

Woods, J.S., Fowler, B.A., 1977b. Renal porphyrinuria during chronic methyl mercury exposure. J. Lab. Clin. Med. 90 (2), 266–272.

Woods, J.S., Fowler, B.A., 1978. Altered regulation of mammalian hepatic heme biosynthesis and urinary porphyrin excretion during prolonged exposure to sodium arsenate. Toxicol. Appl. Pharmacol. 43 (2), 361–371.

Xie, Y., Kondo, M., Koga, H., Miyamoto, H., Chiba, M., 2001. Urinary porphyrins in patients with endemic chronic arsenic poisoning caused by burning coal in China. Environ. Health Prev. Med. 5 (4), 180–185.

Xu, A.S., Ohba, Y., Vida, L., Labotka, R.J., London, R.E., 2000. Aspirin acetylation of betaLys-82 of human hemoglobin. NMR study of acetylated hemoglobin Tsurumai. Biochem. Pharmacol. 60 (7), 917–922.

Yang, R., Pagaduan, J.V., Yu, M., Woolley, A.T., 2014a. On chip preconcentration and fluorescence labeling of model proteins by use of monolithic columns: device fabrication, optimization, and automation. Anal. Bioanal. Chem. 407, 737–747.

Yang, Y., Yang, D., Yang, D., Jia, R., Ding, G., 2014b. Role of reactive oxygen species-mediated endoplasmic reticulum stress in contrast-induced renal tubular cell apoptosis. Nephron. Exp. Nephrol. 128 (1–2), 30–36.

Zang, M., Zhu, F., Zhao, L., Yang, A., Li, X., Liu, H., Xing, J., 2014. The effect of UGTs polymorphism on the auto-induction phase II metabolism-mediated pharmacokinetics of dihydroartemisinin in healthy Chinese subjects after oral administration of a fixed combination of dihydroartemisinin-piperaquine. Malar. J. 13, 478.

Zhang, X., Zhang, X., Niu, Z., Qi, Y., Huang, D., Zhang, Y., 2014. 2,4,6-Trichlorophenol cytotoxicity involves oxidative stress, endoplasmic reticulum stress, and apoptosis. Int. J. Toxicol. 33, 532–541.

Zhang, Y.F., Zhang, M., Huang, X.L., Fu, Y.J., Jiang, Y.H., Bao, L.L., et al., 2015. The combination of arsenic and cryptotanshinone induces apoptosis through induction of endoplasmic reticulum stress-reactive oxygen species in breast cancer cells. Metallomics 7 (1), 165–173.

Zorov, D.B., Juhaszova, M., Sollott, S.J., 2014. Mitochondrial reactive oxygen species (ROS) and ROS-induced ROS release. Physiol. Rev. 94 (3), 909–950.

CHAPTER 3

Computational Toxicology

1 INTRODUCTION

The development of high speed computational tools has been a critical component in the development and acceptance of molecular biomarkers for risk assessment. The reasons for this are manifold. First, computational tools provide a rapid means for both developing relevant storage databases and providing data-mining access to the information in them on an international basis. In addition, since modern analytical instruments equipped with robotics and onboard computer management systems generate gigabytes of data, computer analysis systems are now essential for digesting these outputs and rendering them into a format of value for applying them to risk assessment practice. Looking ahead, it is clear that computational toxicology methods are essential for the further development of systems biology approaches to risk assessment and evaluating the relationships between various classes of molecular biomarkers and adverse outcome pathways leading to cellular toxicity or cancer. A number of these issues have been previously discussed in a prior volume in this series (Please see Fowler, 2013). This chapter will focus in greater detail on some of the more important and evolving aspects of computational toxicology and how these methods have been and may be further used to facilitate public health risk assessments based upon evaluation of molecular toxicology endpoints. This chapter will also examine the ways in which computational toxicology tools may be used to guide laboratory or epidemiology-based research utilizing molecular toxicology derived endpoints. These types of data are needed, in particular, to advance in the field of molecular epidemiology and deal with the problematic issue of disease clusters which are frequently not amenable to useful statistical analysis due to the use of nonspecific or insensitive human health endpoints which require large numbers of subjects in order to detect valid differences. Molecular biomarkers offer an opportunity to address the problem area of disease clusters if properly applied and evaluated since these sensitive biological responses frequently show large response differences relative to standard clinical endpoints. Current challenges to the

CONTENTS

(Continued)

Molecular Biological Markers for Toxicology and Risk Assessment. http://dx.doi.org/10.1016/B978-0-12-809589-8.00003-2

field of molecular biomarker development are validation and interpretation of these responses which are discussed in other chapters of this book. This chapter will attempt to supplement information on computational toxicology presented in first volume of this series (Fowler, 2013) and provide information on how some of these well-developed computational approaches may be used to facilitate molecular biomarker development. There will also be a concerted effort to refer the reader to more detailed references for more specific information since the field of computational biology is also rapidly expanding.

2   DATA MINING APPROACHES—GETTING AN OVERVIEW OF THE CURRENT MOLECULAR BIOMARKER LITERATURE

Over the past 50 years, a number of useful scientific databases have been developed which address a number of elemental aspects essential to credible risk assessments and application of molecular biomarkers to enhance the precision and specificity of these evaluations. Overall, if one steps back and looks at this body of information, it is reasonable to conclude collectively that there is actually a large body of information on molecular biomarkers or relevant molecular/systems biology data that could be utilized to support biomarker development and thus save both time and money. The challenge is how to effectively extract this basic scientific information from the assembled databases in order to facilitate biomarker development with health risk assessments as the ultimate goal. Evolving data mining computer programs are emerging as very powerful tools for this purpose and have been applied to both the peer-reviewed scientific literature and public health databases. Search engines such as Google Scholar are now widely used to rapidly access publications in both peer-reviewed journals and unpublished reports from the "gray literature" which are frequently archived in reports at State/Federal agencies or private companies. Such data may be of value for assessing exposures to chemicals and pharmaceuticals in relation to biomarker development. The overall point is that this information may be accessible and could be helpful for risk assessment purposes if one knows where to look.

3   SOME USEFUL AND PUBLICALLY AVAILABLE DATA RESOURCES

In order to effectively develop new molecular biomarkers for risk assessment, it is important to be aware of the state of the field of risk assessment practice in order to avoid redundancies and understand the value of molecular biomarker approaches for improving needed risk assessments. This requires developing an understanding of available risk assessment resources on an international basis since this field is growing rapidly. The list of useful major risk assessment resources given in following sections is intended as a starting point for new investigators but is not comprehensive since new risk assessment center data resources are rapidly evolving in both the public and private sectors each year. The following section is intended to help the reader gain some insights into some of available databases and *where* to search for information on chemical exposures and toxicity and cancer endpoints.

3.1   United States Agency Public Databases

3.1.1   NIH PubMed Database

The National Library of Medicine supports the well-curated Medline, ToxLine, and PubMed databases. This is an extremely valuable and well-curated source for accessing data from the peer-reviewed literature (Wexler, 2008; Wexler et al., 2012; Wullenweber et al., 2008).

3.1.2   NHANES

The NHANES database, which is managed by the Centers for Disease Control and Prevention, is a large database of excellent analytical and clinical data on representative of the general US population from the age of approximately 2–80 years (Feskanich et al., 1989; Gunter, 1997). It also contains some data sets on some minority populations. The NHANES contains information and banked samples useful for analysis of a number of common toxic chemicals going back approximately 40 years. It is hence a potentially useful data resource for longitudinal studies and trends of chemical exposures for the general US population.

3.1.3   USFDA

The USFDA is a regulatory agency which collects data on the safety of both food and pharmaceutical products. Many of the data collected are proprietary in nature but agency scientists do conduct research and publish information and reports of public health importance (Arvidson, 2008; Arvidson et al., 2010).

3.1.4   USGS

The US Geological Survey (USGS) collects analytical data on a number of chemical agents found in air, soil, and water systems and maintains this information in a large database. It is a potentially rich data source for environmental exposure information which could be linked to human health data from the NHANES, if those data were accessible, since both data sets are generated on a county level basis (Garcia et al., 2001; Putila and Guo, 2011).

3.1.5   ATSDR

This agency produces the highly valuable and widely used ToxProfile series on several hundred important and commonly encountered chemical agents. There is also a smaller set of chemical interaction tox profiles on some commonly encountered mixtures of chemicals. These reports include some evaluation of current biomarker endpoints and the application of these endpoints for risk assessment purposes. The importance of these well- curated ToxProfile documents, which are updated on a regular basis, is that they provide both a relatively current rigorous review of the toxicology literature which is linked to risk assessment evaluations that are peer-reviewed by an interagency committee composed of knowledgeable scientists from other agencies. The ToxProfiles are hence an excellent resource for anyone wishing to quickly obtain an understanding of the current risk assessment status on commonly encountered chemicals or mixtures of chemicals (Duncan and Orr, 2010; Fay and Mumtaz, 1996; Friede and O'Carroll, 1996; Patterson et al., 2002; Sexton et al., 1994; Woodall and Goldberg, 2008).

3.1.6   *USEPA IRIS*

The USEPA Integrated Risk Assessment Information System (IRIS) database is another extensive resource which is extensively used for supporting risk assessment judgments by the EPA. This system is also updated on a regular basis and as such is another valuable source of information on exposures and biological responses to exposures. The risk assessment evaluations derived from the IRIS database are frequently peer reviewed by committees of the NAS/NRC due to their impact on regulatory decision making (Gehlhaus et al., 2011; Ginsberg and Rice, 2005; Guyton et al., 2014; Mortensen and Euling, 2013).

3.1.7   *National Toxicology Program*

The National Toxicology Program (NTP) is an interagency toxicology testing program which is a unit of the National Institute of Environmental Health Sciences (NIEHS) an NIH institute based in Research Triangle Park, North Carolina. The NTP conducts toxicology testing studies on behalf of a number of Federal agencies including the USFDA, USEPA, Consumer Product Safety Commission (CPSC), CDC, ATSDR, and NIOSH. In this role, it also participates in the development of new biomarker tests derived from experimental systems and risk assessment interpretation of data in animal models. This program is hence a rich source of basic toxicology data which could be mined to provide a scientific underpinning for molecular biomarker tests of direct applicability to human health issues derived from analogous chemical exposure situations (Lee et al., 2011; National Toxicology Program, 2012).

4   INTERNATIONAL PUBLIC HEALTH DATABASES

4.1   International Agency for Research Against Cancer

International Agency for Research Against Cancer (IARC) is a component of the WHO which is located in Lyon, France and focuses on a variety of aspects of cancer including research on new methods for early detection and mechanisms of cancer. In this regard, it has a strong interest in molecular biomarkers related to cancer. It is also a major resource for information on cancer which is regularly updated through the IARC Monograph Series. Some examples of IARC publications are as follows: IARC Working Group (2012), Gonzalez-Horta et al. (2015), Herceg et al. (2013), Pearce et al. (2015), Straif et al. (2009). Contact information of the IARC is a follows:

IARC, 150 Cours Albert Thomas, 69372 Lyon CEDEX 08, France, Tel: +33 (0)4 72 73 84 85, Fax: +33 (0)4 72 73 85 75

4.2   The Joint FAO/WHO Expert Committee on Food Additives and WHO Chemical Safety Programs

The United Nations Food and Agriculture Organization (FAO) headquartered in Rome at the address given in next section and The World Health Organization (WHO) with headquarters in Geneva, Switzerland cosponsor the Joint FAO/WHO Expert Committee on Food Additives (JECFA). This is a committee of experts under a global public health umbrella which maintains databases on chemicals in food stuffs and conducts risk assessments on acceptable daily intakes (ADIs). In addition, the International Programme on Chemical Safety of the WHO produces chemical risk assessment reports and follows outbreaks of adverse human health events from chemical exposures such as the large scale arsenic poisonings in India, Pakistan, and Bangladesh related to geological contamination of well-water. The information in these reports is hence a rich resource for linking well-documented chemical exposures to clinical toxicities such as hepatoxicity (Guha Mazumder and Dasgupta, 2011; Islam et al., 2011) and cancers of the skin (Hashim and Boffetta, 2014; Hong et al., 2014; Hunt et al., 2014; Majumdar et al., 2014; Surdu, 2014) and kidneys (Hashim and Boffetta, 2014; Chen, 2014; Saint-Jacques et al., 2014). These data provide excellent background data for linking molecular biomarkers such as arsenic porphyrin excretion patterns to clinical outcomes and conducting rigorous risk assessments. The food safety unit of the FAO may be reached at the contact information given in next section.

FAO is a component of the United Nations which is based in Rome, Italy and provides information on food safety and chemical exposures in a variety of food products. It is hence a potentially useful resource for chemical exposures via food. It may be reached via the following contact information.

4.3   Food Safety and Quality Unit (AGND)

Food and Agriculture Organization of the United Nations
Viale delle Terme di Caracalla
00153 Rome
Italy
Email: food-quality@fao.org
Fax: +39 06 570 54593

4.4   International Program on Chemical Safety

International Program on Chemical Safety (IPCS) is a unit of the WHO which deals with chemical risk assessments. It has a strong interest in the early detection of chemical toxicity and for this reason the field of molecular biomarkers. It is located in the WHO Headquarters in Geneva, Switzerland and may be reached at the following address:

> World Health Organization
> Avenue Appia 20
> 1211 Geneva 27
> Switzerland
> Telephone: +41 22 791 21 11
> Facsimile (fax): +41 22 791 31 11

5   EUROPEAN UNION

In addition to the rich data sources noted earlier, there are a number of repositories of analytical and risk assessment information which have been developed under the auspices of the European Union (EU). The two main EU agencies which are actively engaged in reviewing the scientific literature and conducting risk assessment are the European Food Safety Authority (EFSA) and the European Chemical Agency (ECHA). These agencies monitor chemicals in air, food, and water and conduct risk assessments for major classes of chemicals. In addition, the EU has initiated the REACH initiative (Blaauboer and Andersen, 2007; Greim, 2007; Lewis et al., 2007; Williams et al., 2009) under the auspices of ECHA. More complete descriptions of both EFSA and ECHA along with contact information are provided in following sections. EU requires testing of chemicals used or sold in EU member countries. This is a far-reaching effort which can be expected to further expand with expansion of the "precautionary principle" (PP) and the inclusion of evolving molecular biomarker tests which have increased sensitivity but need further validation studies (Gonzales et al., 2014; Herman and Raybould, 2014; Warshaw, 2012). Nonetheless, the REACH initiative coupled with the PP and the continued evolution of molecular biomarker tests assures that new molecular approaches to chemical risk assessment will come into practice in the EU in coming decades. Some examples of risk assessment practice by EU agencies are reviewed in following sections.

5.1   EFSA

EFSA is based in Parma, Italy and is charged with protecting the safety of the European food supply. It collects information on chemicals in food and animal feed stuffs and has a number of scientific committees which review these data and make risk assessment evaluations. EFSA may be reached using the following contact information.

European Food Safety Authority, Largo N. Palli 5/a, I-43121, Parma
Tel.: +39 0521036111, Fax: +39 0521036110, www.efsa.europa.eu

5.2 European Chemicals Agency (ECHA)

ECHA is another component of the EU which is responsible for monitoring
and regulating chemicals in EU countries through the REACH program. In this
capacity, ECHA has developed a number of read across approaches for evaluating the likely relative toxicity of a number of chemicals based upon QSAR techniques. This agency has a "QSAR Tool Box" located on its website which may
be downloaded to assist in the development of risk assessments. The agency
has its Headquarter in Helsinki, Finland and may be reached via the following
contact information.

Mailing address
   P.O.Box 400
   00121 Helsinki
   Finland

Visiting address
   Annankatu 18
   00120 Helsinki
   Finland

For information on specific questions concerning REACH, CLP, PIC, or the
BPR, the ECHA Helpdesk may be reached at: http://echa.europa.eu/contact.

The ECHA Switchboard may be reached at: +358-9-686180

5.3 Organization for Economic Co-Operation and Development (OECD)

The OECD is a 34-nation group (which includes the United States) founded
in 1948 to help manage the Marshall Plan in Europe after WWII but which
has taken the management of chemicals into its mission. The OECD is based
in Paris, France and collects chemical data and recommends chemical testing
methods. It is an important data source for a number of chemicals. The OECD
may be reached at the following contact information:

   OECD Headquarters
   2, rue André Pascal
   75775 Paris Cedex 16
   France
   Tel.: +33 1 45 24 82 00
   Fax: +33 1 45 24 85 00
   Email: webmaster@oecd.org

There are also OECD regional offices in Berlin, Mexico City, Tokyo, and Washington, DC.

6   CHEMICAL RISK ASSESSMENT RESOURCES IN SELECTED COUNTRIES AND STATES

In addition to the international umbrella organizations noted earlier, it is important to note that specific countries and States within the United States also have active chemical risk assessment programs and maintain databases that may be searched for useful information. It is not possible to provide a comprehensive list of all these programs but the following list, derived from the WHO/IPCS website, are members of the WHO IPCS Risk Assessment Network, and among the most active and potentially helpful resources. One potential benefit in being aware of these agencies as a data resource rests with the ongoing movement of industrial chemical activities to developing countries and hence the increased risk of human exposures. As noted later, the World Health Organization (IPCS) also has a number of cooperating Centers/Units/NGOs engaged in risk assessment which together form a risk assessment network which may be valuable data resources for information on "in-country" chemical exposures in relation to molecular biomarker–based risk assessments. The list given in next section is from the WHO internet site.

7   WHO CHEMICAL RISK ASSESSMENT NETWORK

List of participating WHO Chemical Risk Assessment Network Agencies and Organizations (List reproduced from the World Health Organization, http://who.int/ipcs)

Aarhus University, Department of Environmental Science, Denmark
ANSES, French Agency for Food, Environmental and Occupational Health & Safety, France
ELIKA, Basque Foundation for Agrofood Safety, Spain
Environmental Protection Agency, Ghana
European Food Safety Authority (EFSA), European Union agency
Existing Substances Risk Assessment Program, Health Canada, Canada
Federal Commission for the Protection against Sanitary Risks (COFEPRIS), Mexico
Federal Office of Public Health, Switzerland
Federal Scientific Center for Medical and Preventive Health Risk Management Technologies, Russian Federation
French Society for Environmental Health (SFSE), France
German Federal Institute for Risk Assessment (BfR), Germany
Ghana Heath Service, Ghana
Human Health Risk Assessment Laboratory, Kazakh National Medical University, Kazakhstan

Institut national de santé publique du Québec, Canada

Institute for Risk Analysis and Risk Communication (IRARC), University of Washington, USA

Institute of Chemicals Safety, Chinese Academy of Inspection and Quarantine, China

Institute of Environmental Engineering, Kaunas University of Technology, Lithuania

Institute of Environmental Medicine, Karolinska Institutet, Sweden

Institute of Population Health, University of Ottawa, Canada

Instituto Nacional de Vigilancia de Medicamentos y Alimentos (INVIMA), Colombia

International Centre for Pesticides and Health Risk Prevention (ICPS), Italy

MAK Commission, Germany

Malaysian Society of Toxicology, Malaysia

Ministry of Health, Suriname

National Institute for Public Health and the Environment (RIVM), The Netherlands

National Institute of Food and Drug Safety Evaluation, Republic of Korea

National Institute of Health, Environmental Health Unit, Portugal

Oak Ridge National Laboratory (ORNL), US Department of Energy, USA

Oklahoma Christian University, USA

Pest Management Regulatory Agency, Canada

Pesticides Control Authority, Jamaica

Portuguese Toxicology Association, Portugal

Public Health England, United Kingdom

Research Centre for Toxic Compounds in the Environment (RECETOX), Czech Republic

Royal Society of Chemistry, United Kingdom

Society of Environmental Toxicology and Chemistry (SETAC), USA

Swedish Chemicals Agency (KEMI), Sweden

Swedish Toxicology Sciences Research Center (Swetox), Sweden

Swiss Centre for Applied Human Toxicology, Switzerland

The Institute for Environmental Modeling, University of Tennessee, USA

United States Environmental Protection Agency (EPA), USA

University of A Coruña, Spain

University of North Texas Health Science Center, USA

US Department of Defense, Science and Technology Directorate, USA

WHO Collaborating Centre for Capacity Building and Research in Environmental Health Science and Toxicology at the Chulabhorn Research Institute, Thailand

WHO Collaborating Centre for Environmental Health Sciences at the National Institute of Environmental Health Sciences (NIEHS), USA

WHO Collaborating Centre for Occupational and Environmental
Epidemiology and Toxicology at IRET, Costa Rica
WHO Collaborating Centre for Occupational Health, Finnish Institute of
Occupational Health, Finland
WHO Collaborating Centre for Occupational Health, International
Centre for Rural Health, Italy
WHO Collaborating Centre for Immunotoxicology and Allergic
Hypersensitivity, The Netherlands
WHO Collaborating Centre for Occupational Health at the National
Institute for Occupational Health (NIOH), South Africa
WHO Collaborating Centre on Water and Indoor Air Quality and Food
Safety at NSF International, USA

Nongovernmental organizations in official relations with WHO
Croplife International, Belgium (HQ)
European Centre for Ecotoxicology and Toxicology of Chemicals
(ECETOC), Belgium (HQ)
International Life Sciences Institute (ILSI), USA
International Union of Toxicology (IUTOX), USA

Observer agencies

European Chemicals Agency (ECHA), European Union agency, Finland
Organisation for Economic Co-operation and Development (OECD),
France (HQ)

8 COMPUTATIONAL TOXICOLOGY APPROACHES TO BIOMARKER DEVELOPMENT AND VALIDATION

It should be clear from the previous sections that there are extensive national/international resources being applied to chemical risk assessments and there are established protocols for conducting risk assessments (WHO Risk Assessment Tool Box). While these tools and protocols have been accepted over time, there is general agreement that a pressing need exists to incorporate more modern approaches based on merging advances in analytical chemistry/chemical exposure analysis (eg, LC–MS, MS–MS), with molecular biology (eg, omic biomarkers), and computational toxicology modeling techniques (eg, QSAR, PBPK). The overall purpose is to provide more sensitive and comprehensive data to inform better risk assessment–based decision making. Such information is critical for addressing emerging public health concerns regarding sensitive subpopulations, new classes of chemicals (eg, nanomaterials) for which relatively few toxicological data exist, and the need for expansion of risk assessments across species to support ecological risk assessments.

8.1 US Environmental Protection Agency

The US Environmental Protection Agency (EPA) has strong interest in the development of molecular biomarkers for risk assessment purposes. It is the home of the NEXGEN and TOXCAST programs which are described next and is hence a major resource for persons interested in utilizing molecular biomarkers for risk assessment.

8.1.1 EPA NEXGEN Initiative

The US EPA has taken a leadership role in attempting to incorporate the tools of modern science into various aspects of the chemical risk assessment process. The EPA NEXGEN (Next Generation of Risk Assessment) Initiative is a major program which has several components.

1. Computational Toxicology: This is a large program, based in the Computational Toxicology Center located in Research Triangle Park, North Carolina and conducts a variety of in silico modeling activities
2. ToxCast: This is a high throughput chemical screening program which uses automated in vitro test systems to examine toxic responses of cell lines to selected chemical agents and their metabolites. A main purpose of this program is to develop basic toxicology data on the large number of chemicals in commerce (80,000+) for which few if any toxicology data exist.

8.1.2 Interagency Tox 21 Toxicology Testing Program

The Tox 21 Toxicology Testing Program is an interagency initiative between the National Toxicology Testing Program (NTP) headquartered at NIEHS, with US FDA and the US EPA. The program is intended to decrease the backlog of untested chemicals by leveraging the in vitro testing resources of these agencies with the in vivo testing resources of the NTP.

9 TOXICOLOGY TESTING RESOURCES IN EUROPE

The European Union Joint Research Centre (JRC) and EURL-ECVAM

The JRC and EURL-EVCAM are research units under the European Commission which have ongoing research programs in a number of areas. As with other agencies concerned with public health, there is hence a strong interest in computational toxicology and molecular biomarkers for early detection of toxicity from chemical exposures with attendant reduction in the use of animals for toxicity testing.

Headquarters
European Commission

Environment DG
B - 1049 Brussels
Belgium

The EU Joint Research Centre

This research unit is located in Ispra, Italy. The contact information is:

Centro 13ommune di ricerca
Via Enrico Fermi 2749, I - 21027 Ispra (VA), Italia
Welcome Desk, Bld. 1, TP 018
Main switchboard: +39 0332 789 111
Communication Unit: +39 0332 78 9889
Email: jrc-info@ec.europa.eu

10   COMPUTATIONAL TOOLS FOR CAPTURING THE BIOMARKER LITERATURE

Since the field of biomarkers is expanding very rapidly, it is important to remain current on new developments in this field. Computer-based search engines offer an efficient means to this end. Listed next are the top 10 general search engines as on Jan. 2015 and the most relevant databases of the NIH National Library of Medicine. The value of listing all these excellent resources is to provide a more comprehensive coverage of this evolving area of science.

Ten Most Popular General Search Engines –Jan. 2015 (Source: eBIZ/MBA)

1. Google/Google Scholar
2. Bing
3. Yahoo
4. Ask
5. Aol.
6. Wow
7. WebCrawler
8. MyWebSearch
9. Infospace
10. Info

10.1   National Library of Medicine Databases (www.nlm.nih.gov)

PubMed/MEDLINE
MedlinePlus
ToxNet

11 COMPUTATIONAL APPROACHES FOR ASSISTING IN MOLECULAR BIOMARKER DEVELOPMENT

11.1 Bioinformatic Algorithm Programs for Possible Biomarker Identification

The field of computational bioinformatics is expanded in concert with the field of molecular biology. A major reason for this expansion has been the need to handle and interpret the large volumes of data generated by modern molecular biology techniques. In addition, bioinformatics can be used for identification of important signaling pathways which can serve as putative biomarkers or candidates for further laboratory research. A good example of this approach would be recent studies using a bioinformatics algorithm called OncoFinder for identification of important signaling pathways related to bladder cancer (Borisov et al., 2014; Lezhnina et al., 2014). The value of these approaches is to save both time and expense in molecular biomarker development by efficiently using already generated data to identify the most potentially fruitful avenues of research.

11.2 Computational Approaches for Analysis of Complex Analytical Data Sets

As noted previously, modern analytical methods generate large quantities of data which must be both stored and analyzed in order to provide useful scientific information of value for public health risk assessments. Computational methods are essential in this regard. Techniques such as 2-dimensional gel electrophoresis (2-DGE) which generate large quantities of data are being continuously refined to permit the accurate identification of ever more novel proteins which may serve as putative new molecular biomarkers. Image analysis programs have been developed to sort through the hundreds of spots commonly observed on 2DGE gels (Fowler et al., 2005, 2008) and calculate relative changes in proteomic expression patterns in response to chemical exposures (Fowler et al., 2008). Similar data analysis issues exist for LS/MS or MS/MS data which involve both intact molecules and fragments of proteins generated in response to disease states chemical or drug exposures (Betancourt et al., 2014; Dewalque et al., 2014; Huo et al., 2015; Miller and Spellman, 2014; Rebecca et al., 2014; Sui et al., 2015).

The value of being able to analyze and interpret these data for biomarker development and validation cannot be understated since the risk assessment value of these data rests on their reliability.

11.3 PBPK/SAR/QSAR

Computational modeling approaches such as PBPK, SAR, and QSAR (Demchuk et al., 2011; Ruiz et al., 2010a) have great potential value in expediting molecular biomarker development by documenting the disposition pathways of chemicals and their metabolites in persons as a function of chemical structure, age, gender, and genetic inheritance. This information is very valuable for identifying key metabolic early response pathways to chemical agents as putative biomarker candidates and providing information on possible sensitive populations on the basis of metabolic profiles (eg, high metabolizers versus low metabolizers). In practical terms, PBPK data may be used to determine optimal sampling times for biological markers of exposure to agents such as styrene (Verner et al., 2012) or pesticides such as chlorpyrifos (Mosquin et al., 2009) or carbaryl (Phillips et al., 2014). PBPK/PBTK modeling techniques are also very valuable for reverse dosimetry studies (Grulke et al., 2013; Ruiz et al., 2010a,b).

These types of data are of potentially great value in focusing biomarker development studies into productive areas of research. This approach is also particularly valuable for developing preliminary data on new or relatively unstudied chemical agents and identifying populations at special risk for toxicity prior to the initiation of more costly laboratory-based studies. This "look before you leap" approach using computer modeling techniques can greatly expedite development of effective biomarker tests by providing information on in vivo handling of chemicals to elucidate the optimal time windows following exposures to look for biomarker changes. An example of this approach is the study by Ruiz et al. (2010a) which involved mining the NHANES database for and identifying females in the 6–11 age group as being at special risk for increased cadmium excretion. Urine samples from persons in this age group as could be tested for increased excretion of proteins such as Kim-1 protein, which is a putative biomarker of renal injury (Bailly et al., 2002; Han et al., 2008; Prozialeck and Edwards, 2010, 2012; Prozialeck et al., 2009a,b; Shao et al., 2014; Zhang et al., 2007), to determine if a linkage and dose–response relationship could be established between the observed increases in urinary cadmium and kidney toxicity for the general US population in this age range. If such a relationship could be established, the validity of urinary excretion of Kim-1 as a reliable marker for cadmium-induced kidney toxicity at environmental exposure levels would be expedited. In other words, the computational modeling information is a potentially valuable tool for focusing and guiding molecular biomarker studies.

11.4 PBPK

PBPK modeling approaches have been used for a number of years to model the metabolism and disposition of a variety of drugs and chemicals. More recently, this approach has been applied to evaluating metabolic pathways and helping to identify key steps in the in vivo handling of chemicals to help identify possible biomarkers and to provide a translational bridge between chemical exposures as monitored by the NHANES and biological responses

(Ruiz et al., 2010b, 2015). Such data are of clear importance for validating the sensitivity and specificity of biomarker response profiles in relation to key metabolic steps in the in vivo handling of chemicals such as aromatic solvents (Marchand et al., 2015) or acrylamide (DeWoskin et al., 2013).

11.5 SAR/QSAR

Computational SAR/QSAR techniques have a number of potential roles to play in the development of molecular biomarkers and their application to risk assessment practice (Demchuk et al., 2011). These approaches may help delineate relationships between specific chemicals or their metabolites and effector molecules involved in the mechanisms of toxicity (McPhail et al., 2012; Tie et al., 2012; Xu et al., 2014; Konstantinidou and Hadjipavlou-Litina, 2013). A clear mechanistic understanding of the nature of interactions between a drug or chemical agent and a molecular pathway which plays a central role in their in vivo handling is essential for a correct interpretation of attendant biological responses and accurate risk assessment predications of future health outcomes based upon an understanding of mechanisms of action (Chen et al., 2014; Jung et al., 2013; Lynch et al., 2013; Osgood et al., 2014).

11.6 Molecular Docking Approaches

As noted earlier, SAR/QSAR techniques have broad applicability in explaining mechanisms of toxicity at molecular level. The techniques may also be used as screening tools to identify likely protein targets for xenobiotic chemicals such as POPs which have structural components similar to endogenous chemicals such as steroidal hormones and may hence act as endocrine disruptor agents leading to clinical diseases such as Type II diabetes (Kim et al., 2014; Lee, 2012; Lee et al., 2006, 2007, 2010, 2014; Taylor et al., 2013). The point is that these molecular docking data could be used to explain the associations between specific POPs and hence define a molecular biomarker by delineating the specific component of the POP that is critical for conferring endocrine disrupting capacity by permitting it to dock in the endocrine effector molecule binding site and altering its normal biological activity (Ruiz et al., 2015).

11.7 Pathway Analysis

It is increasingly clear that the normal biology of cells is coordinated by interrelated molecular pathways which regulate the primary biological metabolic machinery of the cells. An initial chemical-induced disturbance of one these pathways may hence have secondary ripple effects in related pathways at the genomic, proteomic, and metabolomics levels (Chen et al., 2015). A widening wave of pathway disorganization due to elevated or prolonged xenobiotic chemical exposure can hence lead to major disruption of normal biology and trigger the activation of adverse outcome pathway leading to cell death (Gustafsson et al., 2014; Liu et al., 2014) or cancer (Atrih et al., 2014; Ciombor et al., 2014; Hu et al., 2014; Kim et al., 2012; Previs et al., 2014; Yang et al., 2014; Yi et al., 2014).

11.8 Systems Biology

Systems biology in global terms is essentially an attempt to gain an overall understanding of how intracellular biological systems interact with each other. In the context of toxicology and molecular biomarker development, this approach is conceptually very useful for examining the myriad of dynamic interactions which occur within a cell and how chemical or pharmaceutical agents may alter these interactions on either a primary or secondary or tertiary basis. This approach may hence lead to a more complete understanding of molecular mechanisms of toxicity and this knowledge may, in turn, be used to elucidate/identify useful molecular biomarkers and their relationships to ongoing processes of cell injury/cell death. It is worth noting that systems biology–based biomarker approaches may also be used to delineate processes such as cell replacement and neoplasia which may frequently cooccur following cell death among a sensitive population of cells. More recently data from metabolomics and genome-wide association studies (GWAS) have been combined to formulate MGWAS approaches that build a systems biology approach which constructs a more comprehensive picture based on the combined strengths of both sets of techniques (Bujak et al., 2014; Demirkan et al., 2015; Dharuri et al., 2014; Komorowsky et al., 2012).

11.9 High Throughput Screening

The rapid ongoing evolution of the various integrated laboratory analytical systems has resulted in the development of high throughput screening platforms for identification of a number of types of putative biomarkers from studies using data from in vitro test systems. A large number of these have been focused on biomarkers for early detection of various cancers (Curtis, 2015; Kim et al., 2015; Lesur et al., 2015; Olsen et al., 2014; Ondrejka and Hsi, 2015; Patel and Ahmed, 2014; Wu et al., 2014) and also various disease entities such as diabetes (Kato and Natarajan, 2014; Marinova et al., 2013), kidney disease (Huang et al., 2014; Salih et al., 2014; Schena, 2014), and skin sensitization (Ayehunie et al., 2009; Forreryd et al., 2014) and the progression of diseases (Ben-Dov et al., 2014; Zeng et al., 2013). The main point of this discussion is that computational techniques are an essential and integrated component for development of molecular biomarkers by virtue of their capacity to digest, analyze, and convert molecular data into a useful format for identification of the most likely biomarker candidates. As noted earlier, this capability saves both time and money by focusing limited laboratory resources on those avenues of research which are most likely to be productive. This is particularly true when coupled with evolving microfluidic analytical approaches (Hughes et al., 2012; Volpetti et al., 2015). Please see Fowler (2013) for a more general review of the various aspects and practical application of these issues.

12 APPLICATIONS OF COMPUTATIONAL METHODS FOR GUIDING BIOLOGICAL MARKER RESEARCH—A SUMMARY

Based on the previous discussion, it seems clear that computational methods already play an important role in development of various types of molecular biomarkers. Among these are the following:

1. Rapid, cost effective, and efficient data mining of the available published literature on biomarkers. This is very important in order to avoid duplication of effort involving expensive laboratory equipment and has contributed to the rapid evolution of this field.
2. Rapid and efficient analysis of data generated by increasingly efficient laboratory systems which generate enormous quantities of data that must be digested in order to be of interpretive value and to advance the field.
3. Generate systems biology analyses such as knowledge maps which delineate molecular relationships that might not otherwise be obvious. This type of graphic information facilitates identification of possible biomarker pathways and also linkages to other pathways which are more directly linked to cell injury/cell death (eg, adverse outcome pathways). This is a very powerful information for predicting health outcomes and is hence a main linkage for the use of molecular biomarkers as credible tools for public health risk assessment.
4. Permit the ready sharing of molecular biomarker findings between laboratories on a global basis via Internet and cloud computing capabilities. Once molecular biomarker data have been analyzed and converted into a common format, they can be more readily shared globally which should permit even more rapid development of the field and application of these modern tools for risk assessments in both developed and developing countries. An excellent case study which will be the focus of the next volume in this series (Fowler, 2017) is the complex issue of "e-waste" which is rapidly growing public health problem which impacts millions of people world-wide due to the large number of chemical agents involved and exposure of persons at early life stages and marginal nutritional status.
5. Finally, the computational tools permit generation of readily understood graphics and facilitate information mapping approaches

for improved risk communication between scientists/risk assessors and societal decision makers who may have limited technical backgrounds and who need to have a basic understanding of complex biomarker-based risk assessment information in order to make sound regulatory or political decisions.

References

Arvidson, K.B., 2008. FDA toxicity databases and real-time data entry. Toxicol. Appl. Pharmacol. 233 (1), 17–19.

Arvidson, K.B., Chanderbhan, R., Muldoon-Jacobs, K., Mayer, J., Ogungbesan, A., 2010. Regulatory use of computational toxicology tools and databases at the United States Food and Drug Administration's Office of Food Additive Safety. Expert Opin. Drug Metab. Toxicol. 6 (7), 793–796.

Atrih, A., Mudaliar, M.A., Zakikhani, P., Lamont, D.J., Huang, J.T., Bray, S.E., et al., 2014. Quantitative proteomics in resected renal cancer tissue for biomarker discovery and profiling. Br. J. Cancer 110 (6), 1622–1633.

Ayehunie, S., Snell, M., Child, M., Klausner, M., 2009. A plasmacytoid dendritic cell (CD123+/CD11c-) based assay system to predict contact allergenicity of chemicals. Toxicology 264 (1–2), 1–9.

Bailly, V., Zhang, Z., Meier, W., Cate, R., Sanicola, M., Bonventre, J.V., 2002. Shedding of kidney injury molecule-1, a putative adhesion protein involved in renal regeneration. J. Biol. Chem. 277 (42), 39739–39748.

Ben-Dov, I.Z., Tan, Y.C., Morozov, P., Wilson, P.D., Rennert, H., Blumenfeld, J.D., Tuschl, T., 2014. Urine microRNA as potential biomarkers of autosomal dominant polycystic kidney disease progression: description of miRNA profiles at baseline. PLoS One 9 (1), e86856.

Betancourt, A., Mobley, J.A., Wang, J., Jenkins, S., Chen, D., Kojima, K., et al., 2014. Alterations in the rat serum proteome induced by prepubertal exposure to bisphenol a and genistein. J. Proteome Res. 13 (3), 1502–1514.

Blaauboer, B.J., Andersen, M.E., 2007. The need for a new toxicity testing and risk analysis paradigm to implement REACH or any other large scale testing initiative. Arch. Toxicol. 81 (5), 385–387.

Borisov, N.M., Terekhanova, N.V., Aliper, A.M., Venkova, L.S., Smirnov, P.Y., Roumiantsev, S., et al., 2014. Signaling pathways activation profiles make better markers of cancer than expression of individual genes. Oncotarget 5 (20), 10198–10205.

Bujak, R., Struck-Lewicka, W., Markuszewski, M.J., Kaliszan, R., 2014. Metabolomics for laboratory diagnostics. J. Pharm. Biomed. Anal. 113, 108–120.

Chen, C.J., 2014. Health hazards and mitigation of chronic poisoning from arsenic in drinking water: Taiwan experiences. Rev. Environ. Health 29 (1–2), 13–19.

Chen, M., Bisgin, H., Tong, L., Hong, H., Fang, H., Borlak, J., Tong, W., 2014. Toward predictive models for drug-induced liver injury in humans: are we there yet? Biomark. Med. 8 (2), 201–213.

Chen, H., Zhu, Z., Zhu, Y., Wang, J., Mei, Y., Cheng, Y., 2015. Pathway mapping and development of disease-specific biomarkers: protein-based network biomarkers. J. Cell Mol. Med. 19 (2), 297–314.

Ciombor, K.K., Haraldsdottir, S., Goldberg, R.M., 2014. How can next-generation sequencing (genomics) help us in treating colorectal cancer? Curr. Colorectal Cancer Rep. 10 (4), 372–379.

Curtis, C., 2015. Genomic profiling of breast cancers. Curr. Opin. Obstet. Gynecol. 27 (1), 34–39.

Demchuk, E., Ruiz, P., Chou, S., Fowler, B.A., 2011. SAR/QSAR methods in public health practice. Toxicol. Appl. Pharmacol. 254 (2), 192–197.

Demirkan, A., Henneman, P., Verhoeven, A., Dharuri, H., Amin, N., van Klinken, J.B., et al., 2015. Insight in genome-wide association of metabolite quantitative traits by exome sequence analyses. PLoS Genet. 11 (1), e1004835.

Dewalque, L., Pirard, C., Charlier, C., 2014. Measurement of urinary biomarkers of parabens, benzophenone-3, and phthalates in a Belgian population. Biomed. Res. Int. 2014, 649314.

DeWoskin, R.S., Sweeney, L.M., Teeguarden, J.G., Sams, II, R., Vandenberg, J., 2013. Comparison of PBTK model and biomarker based estimates of the internal dosimetry of acrylamide. Food Chem. Toxicol. 58, 506–521.

Dharuri, H., Demirkan, A., van Klinken, J.B., Mook-Kanamori, D.O., van Duijn, C.M., t Hoen, P.A., Willems van Dijk, K., 2014. Genetics of the human metabolome, what is next? Biochim. Biophys. Acta. 1842 (10), 1923–1931.

Duncan, M.A., Orr, M.F., 2010. Evolving with the times, the new national toxic substance incidents program. J. Med. Toxicol. 6 (4), 461–463.

Fay, R.M., Mumtaz, M.M., 1996. Development of a priority list of chemical mixtures occurring at 1188 hazardous waste sites, using the HazDat database. Food Chem. Toxicol. 34 (11–12), 1163–1165.

Feskanich, D., Sielaff, B.H., Chong, K., Buzzard, I.M., 1989. Computerized collection and analysis of dietary intake information. Comput. Methods Programs Biomed. 30 (1), 47–57.

Forreryd, A., Johansson, H., Albrekt, A.S., Lindstedt, M., 2014. Evaluation of high throughput gene expression platforms using a genomic biomarker signature for prediction of skin sensitization. BMC Genomics 15, 379.

Fowler, B.A. (Ed.), 2013. Computational Toxicology: Applications for Risk Assessment. Elsevier Publishers, Amsterdam, pp. 258.

Fowler, B.A., 2017. Electronic Waste: Toxicology and Public Health Issues. Elsevier Publishers, Amsterdam. (In Preparation)

Fowler, B.A., Conner, E.A., Yamauchi, H., 2005. Metabolomic and proteomic biomarkers for III–V semiconductors: chemical-specific porphyrinurias and proteinurias. Toxicol. Appl. Pharmacol. 206 (2), 121–130.

Fowler, B.A., Conner, E.A., Yamauchi, H., 2008. Proteomic and metabolomic biomarkers for III–V semiconductors: and prospects for application to nano-materials. Toxicol. Appl. Pharmacol. 233 (1), 110–115.

Friede, A., O'Carroll, P.W., 1996. CDC and ATSDR electronic information resources for health officers. J. Public Health Manag. Pract. 2 (3), 10–24.

Garcia, S.S., Ake, C., Clement, B., Huebner, H.J., Donnelly, K.C., Shalat, S.L., 2001. Initial results of environmental monitoring in the Texas Rio Grande Valley. Environ. Int. 26 (7–8), 465–474.

Gehlhaus, III, M.W., Gift, J.S., Hogan, K.A., Kopylev, L., Schlosser, P.M., Kadry, A.R., 2011. Approaches to cancer assessment in EPA's integrated risk information system. Toxicol. Appl. Pharmacol. 254 (2), 170–180.

Ginsberg, G., Rice, D., 2005. The NAS perchlorate review: questions remain about the perchlorate RfD. Environ. Health Perspect. 113 (9), 1117–1119.

Gonzales, J.F., Barnard, N.D., Jenkins, D.J., Lanou, A.J., Davis, B., Saxe, G., Levin, S., 2014. Applying the precautionary principle to nutrition and cancer. J. Am. Coll. Nutr. 33 (3), 239–246.

Gonzalez-Horta, C., Ballinas-Casarrubias, L., Sanchez-Ramirez, B., Ishida, M.C., Gonzalez-Horta, C., Ballinas-Casarrubias, L., Sanchez-Ramirez, B., Ishida, M.C., Barrera-Hernandez, A., Gutierrez-Torres, D., et al., 2015. A concurrent exposure to arsenic and fluoride from drinking water in chihuahua. Mexico. Int. J. Environ. Res. Public Health 12 (5), 4587–4601.

Greim, H., 2007. Toxicological comments to the discussion about REACH (H. Greim, M. Arand, H. Autrup, H.M. Bolt, J. Bridges, E. Dybing, R. Glomot, V. Foa, R. Schulte-Hermann, Arch Toxicol 2006, 80: 121-124). Reply to the letter to the editor: the need for a new toxicity testing and risk analysis paradigm to implement REACH or any other large scale testing initiative, by B. J. Blaauboer and M. E. Andersen (Arch Toxicol 2007, 81; 385-387). Arch. Toxicol. 81 (12), 895–896.

Grulke, C.M., Holm, K., Goldsmith, M.R., Tan, Y.M., 2013. PROcEED: probabilistic reverse dosimetry approaches for estimating exposure distributions. Bioinformation 9 (13), 707–709.

Guha Mazumder, D., Dasgupta, U.B., 2011. Chronic arsenic toxicity: studies in West Bengal. India Kaohsiung J. Med. Sci. 27 (9), 360–370.

Gunter, E.W., 1997. Biological and environmental specimen banking at the Centers for Disease Control and Prevention. Chemosphere 34 (9–10), 1945–1953.

Gustafsson, M., Nestor, C.E., Zhang, H., Barabasi, A.L., Baranzini, S., Brunak, S., et al., 2014. Modules, networks and systems medicine for understanding disease and aiding diagnosis. Genome Med. 6 (10), 82.

Guyton, K.Z., Hogan, K.A., Scott, C.S., Cooper, G.S., Bale, A.S., Kopylev, L., et al., 2014. Human health effects of tetrachloroethylene: key findings and scientific issues. Environ. Health Perspect. 122 (4), 325–334.

Han, W.K., Waikar, S.S., Johnson, A., Betensky, R.A., Dent, C.L., Devarajan, P., Bonventre, J.V., 2008. Urinary biomarkers in the early diagnosis of acute kidney injury. Kidney Int. 73 (7), 863–869.

Hashim, D., Boffetta, P., 2014. Occupational and environmental exposures and cancers in developing countries. Ann. Glob. Health 80 (5), 393–411.

Herceg, Z., Lambert, M.P., van Veldhoven, K., Demetriou, C., Vineis, P., Smith, M.T., et al., 2013. Towards incorporating epigenetic mechanisms into carcinogen identification and evaluation. Carcinogenesis 34 (9), 1955–1967.

Herman, R.A., Raybould, A., 2014. Expert opinion vs. empirical evidence: the precautionary principle applied to GM crops. GM Crops Food 5 (1), 8–10.

Hong, Y.S., Song, K.H., Chung, J.Y., 2014. Health effects of chronic arsenic exposure. J. Prev. Med. Public Health 47 (5), 245–252.

Hu, W., Liu, T., Ivan, C., Sun, Y., Huang, J., Mangala, L.S., et al., 2014. Notch3 pathway alterations in ovarian cancer. Cancer Res. 74 (12), 3282–3293.

Huang, J.X., Blaskovich, M.A., Cooper, M.A., 2014. Cell- and biomarker-based assays for predicting nephrotoxicity. Expert Opin. Drug Metab. Toxicol. 10 (12), 1621–1635.

Hughes, A.J., Lin, R.K., Peehl, D.M., Herr, A.E., 2012. Microfluidic integration for automated targeted proteomic assays. Proc. Natl. Acad. Sci. USA 109 (16), 5972–5977.

Hunt, K.M., Srivastava, R.K., Elmets, C.A., Athar, M., 2014. The mechanistic basis of arsenicosis: pathogenesis of skin cancer. Cancer Lett. 354 (2), 211–219.

Huo, T., Xiong, Z., Lu, X., Cai, S., 2015. Metabonomic study of biochemical changes in urinary of type 2 diabetes mellitus patients after the treatment of sulfonylurea antidiabetic drugs based on ultra-performance liquid chromatography/mass spectrometry. Biomed. Chromatogr. 29 (1), 115–122.

IARC Working Group on the Evaluation of Carcinogenic Risks to Humans, 2012. Arsenic, metals, fibres, and dusts. IARC Monogr. Eval. Carcinog. Risks Hum. 100 (Pt. C), 11–465.

Islam, K., Haque, A., Karim, R., Fajol, A., Hossain, E., Salam, K.A., et al., 2011. Dose-response relationship between arsenic exposure and the serum enzymes for liver function tests in the individuals exposed to arsenic: a cross sectional study in Bangladesh. Environ. Health 10, 64.

Jung, S.H., Suh, J.H., Kim, E.H., Kim, J.T., Yoo, S.E., Kang, N.S., 2013. The discovery of inhibitors of Fas-mediated cell death pathway using the combined computational method. Bioorg. Med. Chem. Lett. 23 (18), 5155–5164.

Kato, M., Natarajan, R., 2014. Diabetic nephropathy—emerging epigenetic mechanisms. Nat. Rev. Nephrol. 10 (9), 517–530.

Kim, S., Kon, M., DeLisi, C., 2012. Pathway-based classification of cancer subtypes. Biol. Direct. 7, 21.

Kim, K.S., Lee, Y.M., Kim, S.G., Lee, I.K., Lee, H.J., Kim, J.H., et al., 2014. Associations of organochlorine pesticides and polychlorinated biphenyls in visceral vs. subcutaneous adipose tissue with type 2 diabetes and insulin resistance. Chemosphere 94, 151–157.

Kim, Y.J., Sertamo, K., Pierrard, M.A., Mesmin, C., Kim, S.Y., Schlesser, M., et al., 2015. Verification of the biomarker candidates for non-small-cell lung cancer using a targeted proteomics approach. J. Proteome Res. 14, 1412–1419.

Komorowsky, C.V., Brosius, III, F.C., Pennathur, S., Kretzler, M., 2012. Perspectives on systems biology applications in diabetic kidney disease. J. Cardiovasc. Transl. Res. 5 (4), 491–508.

Konstantinidou, M., Hadjipavlou-Litina, D., 2013. Cytokines in terms of QSAR. Review, evaluation and comparative studies. SAR QSAR Environ. Res. 24 (11), 883–962.

Lee, D.H., 2012. Persistent organic pollutants and obesity-related metabolic dysfunction: focusing on type 2 diabetes. Epidemiol. Health 34, e2012002.

Lee, D.H., Lee, I.K., Song, K., Steffes, M., Toscano, W., Baker, B.A., Jacobs, Jr., D.R., 2006. A strong dose-response relation between serum concentrations of persistent organic pollutants and diabetes: results from the National Health and Examination Survey 1999–2002. Diabetes Care 29 (7), 1638–1644.

Lee, D.H., Lee, I.K., Jin, S.H., Steffes, M., Jacobs, Jr., D.R., 2007. Association between serum concentrations of persistent organic pollutants and insulin resistance among nondiabetic adults: results from the National Health and Nutrition Examination Survey 1999–2002. Diabetes Care 30 (3), 622–628.

Lee, D.H., Steffes, M.W., Sjodin, A., Jones, R.S., Needham, L.L., Jacobs, Jr., D.R., 2010. Low dose of some persistent organic pollutants predicts type 2 diabetes: a nested case-control study. Environ. Health Perspect. 118 (9), 1235–1242.

Lee, S.C., Poptani, H., Delikatny, E.J., Pickup, S., Nelson, D.S., Schuster, S.J., et al., 2011. NMR metabolic and physiological markers of therapeutic response. Adv. Exp. Med. Biol. 701, 129–135.

Lee, D.H., Porta, M., Jacobs, Jr., D.R., Vandenberg, L.N., 2014. Chlorinated persistent organic pollutants, obesity, and type 2 diabetes. Endocr. Rev. 35 (4), 557–601.

Lesur, A., Ancheva, L., Kim, Y.J., Berchem, G., van Oostrum, J., Domon, B., 2015. Screening protein isoforms predictive for cancer using immuno-affinity capture and fast LC-MS in PRM mode. Proteomics Clin. Appl. 9, 695–705.

Lewis, A., Kazantzis, N., Fishtik, I., Wilcox, J., 2007. Integrating process safety with molecular modeling-based risk assessment of chemicals within the REACH regulatory framework: benefits and future challenges. J. Hazard Mater. 142 (3), 592–602.

Lezhnina, K., Kovalchuk, O., Zhavoronkov, A.A., Korzinkin, M.B., Zabolotneva, A.A., Shegay, P.V., et al., 2014. Novel robust biomarkers for human bladder cancer based on activation of intracellular signaling pathways. Oncotarget 5 (19), 9022–9032.

Liu, J., Krautzberger, A.M., Sui, S.H., Hofmann, O.M., Chen, Y., Baetscher, M., et al., 2014. Cell-specific translational profiling in acute kidney injury. J. Clin. Invest. 124 (3), 1242–1254.

Lynch, J.T., Cockerill, M.J., Hitchin, J.R., Wiseman, D.H., Somervaille, T.C., 2013. CD86 expression as a surrogate cellular biomarker for pharmacological inhibition of the histone demethylase lysine-specific demethylase 1. Anal. Biochem. 442 (1), 104–106.

Majumdar, K.K., Ghose, A., Ghose, N., Biswas, A., Mazumder, D.N., 2014. Effect of safe water on arsenicosis: a follow-up study. J. Family Med. Prim. Care 3 (2), 124–128.

Marchand, A., Aranda-Rodriguez, R., Tardif, R., Nong, A., Haddad, S., 2015. Human inhalation exposures to toluene, ethylbenzene and m-xylene and physiologically based pharmacokinetic modeling of exposure biomarkers in exhaled air, blood, and urine. Toxicol. Sci. 144, 414–424.

Marinova, M., Altinier, S., Caldini, A., Passerini, G., Pizzagalli, G., Brogi, M., et al., 2013. Multi-center evaluation of hemoglobin A1c assay on capillary electrophoresis. Clin. Chim. Acta 424, 207–211.

McPhail, B., Tie, Y., Hong, H., Pearce, B.A., Schnackenberg, L.K., Ge, W., et al., 2012. Modeling chemical interaction profiles: I. Spectral data-activity relationship and structure-activity relationship models for inhibitors and non-inhibitors of cytochrome P450 CYP3A4 and CYP2D6 isozymes. Molecules 17 (3), 3383–3406.

Miller, R.A., Spellman, D.S., 2014. Mass spectrometry-based biomarkers in drug development. Adv. Exp. Med. Biol. 806, 341–359.

Mortensen, H.M., Euling, S.Y., 2013. Integrating mechanistic and polymorphism data to characterize human genetic susceptibility for environmental chemical risk assessment in the 21st century. Toxicol. Appl. Pharmacol. 271 (3), 395–404.

Mosquin, P.L., Licata, A.C., Liu, B., Sumner, S.C., Okino, M.S., 2009. Reconstructing exposures from small samples using physiologically based pharmacokinetic models and multiple biomarkers. J. Expo. Sci. Environ. Epidemiol. 19 (3), 284–297.

National Toxicology Program, 2012. NTP monograph on health effects of low-level lead. NTP Monogr. xiii, xv–148.

Olsen, L., Campos, B., Winther, O., Sgroi, D.C., Karger, B.L., Brusic, V., 2014. Tumor antigens as proteogenomic biomarkers in invasive ductal carcinomas. BMC Med. Genomics 7 (Suppl. 3), S2.

Ondrejka, S.L., Hsi, E.D., 2015. Pathology of B-cell lymphomas: diagnosis and biomarker discovery. Cancer Treat. Res. 165, 27–50.

Osgood, R.S., Upham, B.L., Hill, III, T., Helms, K.L., Velmurugan, K., Babica, P., Bauer, A.K., 2014. Polycyclic aromatic hydrocarbon-induced signaling events relevant to inflammation and tumorigenesis in lung cells are dependent on molecular structure. PLoS One 8 (6), e65150.

Patel, S., Ahmed, S., 2014. Emerging field of metabolomics: big promise for cancer biomarker identification and drug discovery. J. Pharm. Biomed. Anal. 107c, 63–74.

Patterson, J., Hakkinen, P.J., Wullenweber, A.E., 2002. Human health risk assessment: selected Internet and world wide web resources. Toxicology 173 (1–2), 123–143.

Pearce, N.E., Blair, A., Vineis, P., Ahrens, W., Andersen, A., Anto, J.M., et al., 2015. IARC monographs: 40 years of evaluating carcinogenic hazards to humans. Environ. Health Perspect. 123, 507–514.

Phillips, M.B., Yoon, M., Young, B., Tan, Y.M., 2014. Analysis of biomarker utility using a PBPK/PD model for carbaryl. Front. Pharmacol. 5, 246.

Previs, R.A., Coleman, R.L., Harris, A.L., Sood, A.K., 2014. Molecular pathways: translational and therapeutic implications of the notch signaling pathway in cancer. Clin. Cancer Res. 21, 955–961.

Prozialeck, W.C., Edwards, J.R., 2010. Early biomarkers of cadmium exposure and nephrotoxicity. Biometals 23 (5), 793–809.

Prozialeck, W.C., Edwards, J.R., 2012. Mechanisms of cadmium-induced proximal tubule injury: new insights with implications for biomonitoring and therapeutic interventions. J. Pharmacol. Exp. Ther. 343 (1), 2–12.

Prozialeck, W.C., Edwards, J.R., Lamar, P.C., Liu, J., Vaidya, V.S., Bonventre, J.V., 2009a. Expression of kidney injury molecule-1 (Kim-1) in relation to necrosis and apoptosis during the early stages of Cd-induced proximal tubule injury. Toxicol. Appl. Pharmacol. 238 (3), 306–314.

Prozialeck, W.C., Edwards, J.R., Vaidya, V.S., Bonventre, J.V., 2009b. Preclinical evaluation of novel urinary biomarkers of cadmium nephrotoxicity. Toxicol. Appl. Pharmacol. 238 (3), 301–305.

Putila, J.J., Guo, N.L., 2011. Association of arsenic exposure with lung cancer incidence rates in the United States. PLoS One 6 (10), e25886.

Rebecca, V.W., Wood, E., Fedorenko, I.V., Paraiso, K.H., Haarberg, H.E., Chen, Y., et al., 2014. Evaluating melanoma drug response and therapeutic escape with quantitative proteomics. Mol. Cell Proteomics 13 (7), 1844–1854.

Ruiz, P., Fowler, B.A., Osterloh, J.D., Fisher, J., Mumtaz, M., 2010a. Physiologically based pharmacokinetic (PBPK) tool kit for environmental pollutants—metals. SAR QSAR Environ. Res. 21 (7–8), 603–618.

Ruiz, P., Mumtaz, M., Osterloh, J., Fisher, J., Fowler, B.A., 2010b. Interpreting NHANES biomonitoring data, cadmium. Toxicol. Lett. 198 (1), 44–48.

Ruiz, P., Perloina, A., Mumtaz, M., Fowler, B.A., 2015. A systems biology approach reveals converging molecular mechanisms that link POPs to common metabolic diseases. Environ. Health Perspect. December 18. (Epub ahead of print)

Saint-Jacques, N., Parker, L., Brown, P., Dummer, T.J., 2014. Arsenic in drinking water and urinary tract cancers: a systematic review of 30 years of epidemiological evidence. Environ. Health 13, 44.

Salih, M., Zietse, R., Hoorn, E.J., 2014. Urinary extracellular vesicles and the kidney: biomarkers and beyond. Am. J. Physiol. Renal. Physiol. 306 (11), F1251–F1259.

Schena, F.P., 2014. Biomarkers and personalized therapy in chronic kidney diseases. Expert Opin. Investig. Drugs 23 (8), 1051–1054.

Sexton, K., Wagener, D.K., Selevan, S.G., Miller, T.O., Lybarger, J.A., 1994. An inventory of human exposure-related data bases. J. Expo. Anal. Environ. Epidemiol. 4 (1), 95–109.

Shao, X., Tian, L., Xu, W., Zhang, Z., Wang, C., Qi, C., et al., 2014. Diagnostic value of urinary kidney injury molecule 1 for acute kidney injury: a meta-analysis. PLoS One 9 (1), e84131.

Straif, K., Benbrahim-Tallaa, L., Baan, R., Grosse, Y., Secretan, B., El Ghissassi, F., et al., 2009. A review of human carcinogens—Part C: metals, arsenic, dusts, and fibres. Lancet Oncol. 10 (5), 453–454.

Sui, W., Zhang, R., Chen, J., He, H., Cui, Z., Ou, M., et al., 2015. Comparative proteomic analysis of membranous nephropathy biopsy tissues using quantitative proteomics. Exp. Ther. Med. 9 (3), 805–810.

Surdu, S., 2014. Non-melanoma skin cancer: occupational risk from UV light and arsenic exposure. Rev. Environ. Health 29 (3), 255–264.

Taylor, K.W., Novak, R.F., Anderson, H.A., Birnbaum, L.S., Blystone, C., Devito, M., et al., 2013. Evaluation of the association between persistent organic pollutants (POPs) and diabetes in epidemiological studies: a national toxicology program workshop review. Environ. Health Perspect. 1217, 774–783.

Tie, Y., McPhail, B., Hong, H., Pearce, B.A., Schnackenberg, L.K., Ge, W., et al., 2012. Modeling chemical interaction profiles: II. Molecular docking, spectral data-activity relationship, and structure-activity relationship models for potent and weak inhibitors of cytochrome P450 CYP3A4 isozyme. Molecules 17 (3), 3407–3460.

Verner, M.A., McDougall, R., Johanson, G., 2012. Using population physiologically based pharmacokinetic modeling to determine optimal sampling times and to interpret biological exposure markers: the example of occupational exposure to styrene. Toxicol. Lett. 213 (2), 299–304.

Volpetti, F., Garcia-Cordero, J., Maerkl, S.J., 2015. A microfluidic platform for high-throughput multiplexed protein quantitation. PLoS One 10 (2), e0117744.

Warshaw, J., 2012. The trend towards implementing the precautionary principle in us regulation of nanomaterials. Dose Response 10 (3), 384–396.

Wexler, P., 2008. Online toxicology resources in support of risk assessment from the U.S. National Library of Medicine. Toxicol. Appl. Pharmacol. 233 (1), 63.

Wexler, P., Gilbert, S.G., Thorp, N., Faustman, E., Breskin, D.D., 2012. The World Library of Toxicology, Chemical Safety, and Environmental Health (WLT). Hum. Exp. Toxicol. 31 (3), 207–214.

Williams, E.S., Panko, J., Paustenbach, D.J., 2009. The European Union's REACH regulation: a review of its history and requirements. Crit. Rev. Toxicol. 39 (7), 553–575.

Woodall, G.M., Goldberg, R.B., 2008. Summary of the workshop on the power of aggregated toxicity data. Toxicol. Appl. Pharmacol. 233 (1), 71–75.

Wu, X., Zahari, M.S., Renuse, S., Jacob, H.K., Sakamuri, S., Singal, M., et al., 2014. A breast cancer cell microarray (CMA) as a rapid method to characterize candidate biomarkers. Cancer Biol. Ther. 15 (12), 1593–1599.

Wullenweber, A., Kroner, O., Kohrman, M., Maier, A., Dourson, M., Rak, A., et al., 2008. Resources for global risk assessment: the International Toxicity Estimates for Risk (ITER) and Risk Information Exchange (RiskIE) databases. Toxicol. Appl. Pharmacol. 233 (1), 45–53.

Xu, L., Zhang, Y., Dai, W., Wang, Y., Jiang, D., Wang, L., et al., 2014. Design, synthesis and SAR study of novel trisubstituted pyrimidine amide derivatives as CCR4 antagonists. Molecules 19 (3), 3539–3551.

Yang, W., Yoshigoe, K., Qin, X., Liu, J.S., Yang, J.Y., Niemierko, A., et al., 2014. Identification of genes and pathways involved in kidney renal clear cell carcinoma. BMC Bioinformatics 15 (Suppl. 17), S2.

Yi, T., Zhai, B., Yu, Y., Kiyotsugu, Y., Raschle, T., Etzkorn, M., et al., 2014. Quantitative phosphoproteomic analysis reveals system-wide signaling pathways downstream of SDF-1/CXCR4 in breast cancer stem cells. Proc. Natl. Acad. Sci. USA 111 (21), E2182–E2190.

Zeng, T., Sun, S.Y., Wang, Y., Zhu, H., Chen, L., 2013. Network biomarkers reveal dysfunctional gene regulations during disease progression. FEBS J. 280 (22), 5682–5695.

Zhang, Z., Humphreys, B.D., Bonventre, J.V., 2007. Shedding of the urinary biomarker kidney injury molecule-1 (KIM-1) is regulated by MAP kinases and juxtamembrane region. J. Am. Soc. Nephrol. 18 (10), 2704–2714.

CHAPTER 4

Omic Biological Markers

1 INTRODUCTION

The advent of modern tools of analytical chemistry, molecular biology, and computational biology in recent decades has resulted in a rapid expansion of basic scientific knowledge about how drugs or chemicals may perturb normal biological processes. The application of this knowledge for risk assessment purposes is a relatively new area of scientific endeavor which has been spearheaded by the leadership of programs such as the USEPA NEXGEN initiative (Cote et al., 2012; Krewski et al., 2014; McConnell et al., 2014; Meek et al., 2014; Simon et al., 2014). This initiative is intended to integrate the tools of modern molecular biology coupled with application of computational methodologies to develop mechanism-based "mode of action risk assessments," which should greatly improve speed, accuracy, and precision of chemical risk assessments going forward in time. This chapter will briefly review some of the more intense areas of ongoing activity such as the "omics" biomarkers and related aspects from the perspective of how these modern tools may be applied to improving public health risk assessments and advancing the science of toxicology through basic mechanistic understandings of how drugs or chemicals interact with biological systems and initiate cell injury processes.

1.1 Genomics

The rapidly expanding field of genomics biomarker development is currently centered around several main areas of activity. These include genome-wide association studies (GWAS), the regulatory roles of microRNAs (mRNAs) in regulating epigenetic processes, single nucleotide polymorphisms (SNPs) in helping to define populations at risk for adverse outcomes, statistical/Bayesian approaches for evaluating putative biomarkers from genomic specific datasets or curated databases and integrated computational approaches for delineating relationships between specific genes, pathways, and molecular factors involved in regulation of these systems. The following discussion will briefly examine each of the previously mentioned areas and focus on how knowledge generated

CONTENTS

(Continued) **63**

Molecular Biological Markers for Toxicology and Risk Assessment. http://dx.doi.org/10.1016/B978-0-12-809589-8.00004-4

by these tools has contributed or could contribute to furthering improvements in public health risk assessments for chemical agents or pharmaceuticals.

1.1.1 GWAS Studies

With the sequencing of the human genome (Bakulski et al., 2015; Braundmeier et al., 2015; Byrd and Segre, 2015; Ilyas et al., 2015; Kloosterman et al., 2015) a large body of genomic information became available to aid scientific research in a number of fields including development of biomarkers for delineating susceptibility (Bakulski et al., 2015), mechanisms of toxicity/cancer (Chappell et al., 2014; Hughes et al., 2015; Kelly and Vineis, 2014; Liao et al., 2015; Pan et al., 2013; Williams et al., 2015; Yin et al., 2015; Zhang et al., 2015b), and systems biology approaches to risk assessment (Bertrand et al., 2015; Bhattacharjee et al., 2013; Combes, 2012; Donovan and Cordon-Cardo, 2014; Ellis et al., 2014; Goodman et al., 2014; Kamitsuji et al., 2015; Kelly and Vineis, 2014; Kleensang et al., 2014; McEwen, 2015; Ngalame et al., 2013; Pan et al., 2013; Panagiotou and Taboureau, 2012; Patel and Cullen, 2012; Schrattenholz et al., 2012). Clearly evaluation of the human genome databases (Kamitsuji et al., 2015; Negi et al., 2015; Smedley et al., 2015; Tseng et al., 2015; Welch et al., 2015; Zou et al., 2015) provide a rich first step opportunity for biomarker development if it is mined effectively in both prospective manner and retrospective manner to complement other more specific areas of investigation.

1.1.2 mRNAs

The regulatory roles of noncoding mRNAs in mediating epigenetic effects such as DNA methylation/gene silencing have important risk assessment implications for defining subpopulations at risk for toxicity (Gim et al., 2014; Papageorgiou et al., 2015).

1.1.3 SNPs

Single nucleotide polymorphisms (SNPs) are single DNA base changes that produce variations in the functionality of gene products which may result in altered susceptibility to toxicity from agents such as lead (Krieg et al., 2009; Neslund-Dudas et al., 2014; Scinicariello and Portier, 2015; Scinicariello et al., 2010; Sobin et al., 2011), arsenic (Isokpehi et al., 2012; Pierce et al., 2012), and organic chemicals (Boso et al., 2014; Ng et al., 2015). From the perspective of risk assessment, these changes may have profound implications for defining subpopulations at special risk for toxicity.

The related field of epigenomics is also rapidly expanding and includes DNA methylation (Bachmann, et al., 2010) and histone modifications (Kouzarides, 2007) as important mechanisms for regulation of gene activity which may be altered by chemical exposures. The enzymatic processes which conduct these processes and the genes which govern them are a productive and complex area of ongoing research. One important aspect of understanding the roles of epigenetics mechanisms in gene regulation is the dynamic nature of these processes which may vary as a function gender, age, and nutritional interactions. Taken together, these factors play important roles in regulating "phenotypic plasticity" which appears to be important in mediating susceptibility to adverse outcomes such as cell death and initiation of carcinogenesis. Such information is hence central to development of mode of action–based risk assessments for chemical exposures.

2 STATISTICAL/BAYESIAN APPROACHES FOR DELINEATING BIOMARKERS

It is clear that advent of high-throughput genomic techniques and assembly of genomic databases has resulted in the need for advanced computational methods for analyzing the large amount of data generated. The application of statistical methods such as ANOVAs (Alonso et al., 2015; Mischak et al., 2015; Ramirez-Santana et al., 2015; Xia et al., 2015) and Bayesian analysis (Akutekwe and Seker, 2014; Graziani et al., 2015; Hack et al., 2010) for evaluating putative biomarkers has proven essential to the advancement of genomic-based biomarkers.

2.1 Applications of Genomic/Epigenomic Biomarkers for Risk Assessment

The field of genomic/epigenomic biomarkers continues to rapidly expand and much has been learned about the initial responses of genes to drugs and chemical agent exposures. In terms of improving risk assessments, this information is valuable in terms of a basic understanding of the roles of genetic inheritance as a function of gender, age and epigenetic factors in mediating susceptibility to toxicity and defining subpopulations at special risk. There is a growing appreciation of need to protect sensitive subgroups and answer the question of "dose makes the poison to whom?" There are profound practical implications of addressing this question correctly since it will impact permissible exposure levels and legal regulatory guidelines. From this perspective, credible genomic information could be invaluable in helping to define persons or life stages at special risk and moving the field of risk assessment away from procedurally derived uncertainty factors (UFs). The UF approach has served the field of chemical risk assessment relative well over the years and in the absence of hard biochemical data but we are now in a new era of systems biology–based molecular information and it would seem unwise not to embrace these new data as they become validated and accepted in the scientific community. As noted earlier, the EPA NEXGEN program is moving forward to incorporate systems biology approaches into risk assessment practice for that agency and while it takes some years in order to reach fruition, there is good reason for optimism in the ultimate success of this approach given the continuing rapid expansion of knowledge in modern genomics.

2.2   Genomic Risk Assessment Case Study

2.2.1   ALAD Polymorphisms and Risk of Lead-Induced Hypertension as an Example

It has been appreciated for a number of years that exposure to lead is associated with development of hypertension (NAS, 1993) which is a major public health problem but metaanalysis studies (Scinicariello et al., 2005) have shown that individual vary greatly in the development of hypertension and other health effects over similar ranges of blood lead concentrations. Delta-aminolevulinic acid dehydratase (ALAD) is a zinc-dependent enzyme which catalyzes the second step in the heme biosynthetic pathway (Goering and Fowler, 1987a, 1987b) is the major carrier protein for lead in blood (Bergdahl et al., 1997) and exists as two distinct alleles (ALAD 1 and ALAD 2) which vary in their ability to bind lead and hence regulate the bioavailability of this metal to other sensitive biological systems such as those regulating blood pressure. ALAD 2 has the high affinity for lead is associated with elevated blood lead values in blood, ALAD is localized in red blood cells as a soluble protein. ALAD 1 is the more common allele (90%) while ALAD 2 is less common (10%) and hybrid (ALAD/ALAD2) carriers also exist since the ALAD enzyme complex occurs as an octomeric protein complex (Dent et al., 1990).

Stratified analyses of the large NHANES cohort which looked at possible relationships between blood lead and hypertension as a function of ALAD genotype (Table 4.1) disclosed that carriers of the ALAD2 or ALAD1/2 genotype had an elevated risk of hypertension at mid–high level blood values found in the general US population (Scinicariello et al., 2010). The practical point here from the risk assessment perspective is that knowledge of this genotypic relationship between a specific ALAD allele and hypertension over a blood lead range common in the general US population provides valuable public health information for defining a population at special risk for lead-induced hypertension in the general US population and guidance developing sound

Table 4.1 Adjusted Odds Ratio (for Race, Sex, Age, Education, Smoking Status, Alcohol Intake, Body Mass Index, Serum Creatinine, Serum Total Calcium, and Glycated Hemoglobin) for Hypertension in ALAD 2 Carriers Stratified by Lead Quartile

Lead Quartile (µg/dL)	ALAD 1-1	ALAD 1-2/2-2
0.7–1.4	1.00	0.79 (0.17–3.74)
1.5–2.3	1.00	1.12 (0.51–2.44)
2.4–3.7	1.00	0.68 (0.39–1.16)
3.8–52.9	1.00	1.83 (1.11–3.03)

ATSDR

molecular biomarker–based risk management strategies. It should be noted that there a number of other major genotyped proteins in the NHANES data set (Chang et al., 2009, 2010; Khoury et al., 2009; Krieg et al., 2010; van Bemmel et al., 2011) which could be evaluated as possible molecular biomarkers to look for possible relationships between genotypic expression and risk of susceptibility to chemical-induced cell injury or development of cancer.

3 PROTEOMICS

Proteomic biomarkers examine changes in translated gene products that are actually synthesized or produced in response to chemical agents such as arsenic (Conner et al., 1993) or physical stimuli (eg, heat or radiation). Changes in protein expression patterns may be followed by use of two-dimensional gel electrophoresis (2-DGE) or more recently by mass spectrometry techniques (Burke et al., 2015; Zhang et al., 2015a; Zhao et al., 2015) or a combination of both techniques. 2-DGE computerized digital image analysis methods (Tables 4.2 and 4.3) may be used to delineate semiquantitative changes in protein expression patterns to assess relative changes in protein expression patterns (See Fowler et al., 2005, 2008). Mass spectroscopy techniques may also be used to quantitatively measure changes in altered proteins or fragments of expressed proteins (Biggar and Li, 2015; Burke et al., 2015; Zhang et al., 2015a) or urinary excretion expressed proteins (Zhao et al., 2015) These common proteomic analytical approaches generate large quantities of data which clearly require computational methods in order extract information of value for risk assessment purposes. This is particularly important for new classes of chemicals or mixtures of chemicals for which little or no prior toxicological information are available.

Newer high technology materials such as the binary (III–V) semiconductors gallium arsenide (GaAs) and indium arsenide (InAs) or microparticulate formulations of these materials are useful examples of this problem area given the global expansion of this industry. The generation of millions of tons of electronic "e-waste" due to the unregulated recycling of older electronic devices such as computer, printers, cell phones, and inexpensive digital clocks and radios in developing countries highlights the need for new public health risk assessment approaches focused on preventing human illnesses from exposure to these materials. Proteomic biomarkers offer one relatively inexpensive approach to early detection of initial target organ effects from low dose exposures.

In vivo exposure studies in hamsters which compared the relative effects GaAs and InAs on protein expression demonstrated clear gender differences in proteomic responses to these two binary compounds (Tables 4.2 and 4.3). These differences in gender-based response patterns were also consistent between hamsters and human kidney cells in primary culture. Males and females had different proteomic responses to the two semiconductor compounds (Fowler et al., 2008). The data in the tables are expressed as changes in the digitized spot intensities from animals treated with GaAs or InAs particles relative to those from control animals. In addition, analysis of urinary proteinuria patterns in hamsters following silver staining demonstrated greater protein excretion in male animals following exposure to InAs relative to GaAs and that this effect was associated with an apparent inhibition of protein synthesis in animals exposed to InAs (Fowler et al., 2005). The value of this information from a risk assessment perspective is that demonstrates gender differences in responses to these semiconductor materials and that inhibition of protein thesis in renal tubule

Table 4.2 Changes in Polypeptides at 30 Days After Exposure to InAs or GaAs in Male Hamster Kidney Cells

MW Range	Comma Spot#[a]	InAs[a]	GaAs[b]
100–90	1	1.00	2.1[c]
	18	1.00	0.7
89–70	2	—	—
	12	—	1.1
69–50	3	90	4.0[c]
	11	—	1.2
	22	—	—
	23	—	—
	25	—	—
	32	—	—
49–40	10	0.85	1.8
	16	1.20	1.6
	26	—	—
	28	—	—
39–30	4	0.79	2.0[c]
	5	1.80	0.5[c]
	6	—	0.1[c]
	7	—	0.3[c]
	8	1.50	3.6[c]
	13	0.90	0.1[c]
	14	0.72	0.6[c]
	15	1.20	1.7
	17	0.74	2.4[c]
	19	—	—
	20	—	—
	30	—	—
29–20	31	—	—

[a]Spot number denoted on gel.
[b]Spot intensity (ratio of control).
[c]a ≥ twofold increase/decrease in polypeptides.

Table 4.3 Changes in Polypeptides at 30 Days After Exposure to InAs or GaAs in Female Hamster Kidney Cell

MW Range	Common Spot#[a]	InAs[a]	GaAs[b]
100–90	1	0.92	0.96
	18	0.56	0.17[c]
	29	0.71	0.37[c]
89–70	2	6.60[c]	6.60[c]
	12	0.47[c]	1.20
69–60	3	0.68	1.10
	11	0.61	0.96
	22	0.32[c]	0.78
	23	0.31[c]	0.83
	24	0.26[c]	0.64
	25	0.26[c]	1.26
	32	0.90	0.96
49–40	10	0.43[c]	0.90
	16	2.50[c]	0.57
	26	0.58	0.45[c]
	27	—	1.60
	28	—	10.00[c]
	33	—	1.40
	34	0.43[c]	0.57
39–30	4	—	—
	5	2.00[c]	1.20
	6	0.14[c]	0.70
	7	0.32[c]	0.61
	8	0.39[c]	0.50
	13	0.47[c]	1.20
	14	0.37[c]	1.00
	15	0.76[c]	1.25
	17	1.00	0.96
	19	0.20[c]	1.10
	20	—	0.50
≤29	30	0.84	0.76
	31	0.54	0.60

[a]Spot number denoted on gel.
[b]Spot intensity (ratio of control).
[c]a ≥ twofold increase/decrease in polypeptides.

cells is associated with increased renal toxicity as judged by increases in urinary protein excretion. In other words, gender and compound-specific inhibition of protein synthesis must be taken into consideration in incorporating results of experimental studies for relatively unstudied chemicals such as the III–V semiconductors or mixtures of such chemicals in to human health risk assessments.

3.1 Applications of Proteomics for Risk Assessment

As with genomics, the related field of proteomics offers a number of opportunities for improving risk assessments by delineating which genes showing altered expression patterns are actually expressed as gene products. In addition, proteomic techniques may also demonstrate posttranslational modifications of proteins and protein–protein interactions which are frequently important in regulating the biological activity of expressed proteins. These types of data are hence potentially very valuable toward understanding mechanisms of toxicity for mode of action–based risk assessments. Proteomic data clearly compliment data from genomic and metabolomics/metabonomic studies and hence contribute to a more holistic appreciation of early/ongoing chemical-induced disturbances in the normal biology of cells. Ultimately such comprehensive datasets provide a foundation for the future development of individual-based risk assessments which incorporate information from all levels of biological organization using systems biology–based approaches.

3.2 Proteomic Risk Assessment Case Study

3.2.1 Differences in Proteomic Expression Patterns and Risk of Toxicity From Exposure to III–V Semiconductor Compounds

It is clear that we live in an increasingly chemical world with many new chemical agents being introduced every year. The semiconductor industry is a useful example of an industrial activity which has contributed many new useful devices made from chemicals of mixtures of or mixtures of chemicals for which only limited toxicological information is available. Application of molecular biomarker approaches to the area of new chemicals and chemical mixtures provides a potentially important new set of tools for assessing possible public health risks to workers engaged in making semiconductor or similar high-technology materials but also those engaged in recycling electronic devices made from these materials at the end of their useful lifetimes. The topic of "e-waste" will be discussed in the third volume of this book series and will include some practical suggestions on how molecular-based biomarkers could be incorporated into risk assessment strategies of value in developing countries where public health resources may be limited.

The following proteomic biomarker examples will compare cellular responses to the binary semiconductor compounds GaAs and InAs at equivalent dose levels following in vivo dosing in hamsters and in vitro hamster or human primary cell cultures as a function of gender. These data are potentially important from risk assessment perspective since the semiconductor industry has had a relatively high percentage of female employees (Lee et al., 2011; Lin et al., 2008).

3.2.2 In Vivo Exposure Studies of Hamsters

A complete description of these studies which compared the relative toxicity of GaAs and InAs particles is provided elsewhere (Conner et al., 1993, 1995) but the primary focus was to evaluate the relative toxicity of these two major III–V semiconductor compounds in relation to each other and in relation to proteomic cellular responses in vivo since relatively little basic toxicology data was available on them. The preliminary studies showed that InAs was relatively more toxic than GaAs as judged by urinary proteinuria patterns and that the observed cellular proteomic response patterns varied as a function of gender (Fowler et al., 2005).

3.2.3 In Vitro Exposure Studies in Both Hamster and Human Primary Cultures

Comparative in vitro studies using hamster and human primary cultures of renal tubule cells demonstrated similar findings with regard to the relative toxicity of Ga, As, and In and proteomic response differences between cells from male and females of each species (Fowler et al., 2008) at equivalent dose levels. The value of this information rests with an appreciation of the need to include gender differences in any risk assessment for these semiconductor materials on an individual or binary mixture basis.

4 METABOLOMICS/METABONOMICS

Metabolomic or metabonomic biomarkers as they are sometimes termed essentially follow chemical- or drug-induced disturbances in metabolic pathways to provide insights into early manifestations of toxicity. In this manner, they provide useful information at another level of biological organization which complements data from genomic and proteomic biomarker studies. The combination of information from these three interrelated classes of biomarkers hence has the potential to provide more holistic (360 degree) assessment of ongoing toxic processes at early stages of development and prior to the onset of overt clinical disease. From a predictive risk assessment perspective, this is potentially very valuable if one can pull together and interpret the assorted pieces of information. As discussed later, computational toxicology methods are the most likely approach for integrating such diverse but biologically interrelated data sets. The following discussion will examine several types of metabolomic/metabonomic biomarkers and look at these types of biomarker responses in relation to complementary genomic or proteomic biomarkers where such data exist.

4.1 Heme Biosynthetic Pathway

The heme biosynthetic pathway is essential for life and produces heme for a number of critical biological processes such as hemoglobin and cytochromes for cellular respiration and drug metabolizing enzyme activities. There are a number of steps in this pathway which have been found to be differentially and highly sensitive to a number of toxic agents such as chlorinated organic chemicals such as PCBs and dioxins (Boyd et al., 1989) and metallic elements

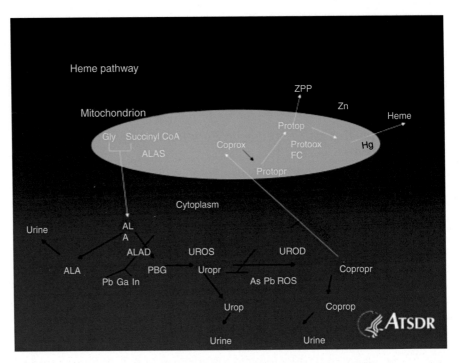

FIGURE 4.1 Global diagram of the heme biosynthetic pathway showing individual steps and points of inhibition for Pb, Hg, Ga, In, and As, which result in element specific disturbances of heme metabolism and increased excretion of heme precursors.

such as lead, mercury, arsenic indium, and gallium (See Fowler et al., 2005, Silbergeld and Fowler, 1987). Differential chemical agent inhibition of specific steps in the heme biosynthetic pathway (Fig. 4.1) has been shown to produce increased urinary excretion of specific porphyrins/heme precursors in agent (Cantoni et al., 1984) or mixture-specific (Conner et al., 1995; Dunagin, 1984; Fowler and Mahaffey, 1978) patterns as discussed later. The value of this information for risk assessment purposes rests with the fact that these metabolic disturbances occur early in the toxic processes and may actually exacerbate chemical agent toxicity since the porphyrins/heme precursors are also toxic by virtue of generating reactive oxygen species (ROS). In other words, disturbance of the heme pathway and resulting consequences hence become a component of the ultimate toxic process and should be factored into any overall risk assessment for chemicals which perturb this pathway on an individual or chemical mixture basis. Clearly, the potential impact of these secondary chemical-induced disturbances in heme and porphyrin metabolism may greatly complicate accurate risk assessments and again, the application of computational methods to account for these effects seem warranted in order to avoid underestimating potential risk. In addition, since there are a number of genetic variants in the enzymes in the heme biosynthetic pathway (Tzou et al., 2014; Wetmur et al., 1991), it is important to take these variations into account with regard to addressing the issue of sensitive subpopulations at special risk for toxicity.

4.2 Applications of Metabolomics/Metabonomics for Risk Assessment—the Heme Biosynthetic Pathway as an Example

Metabolomic/metabonomic data provide a third tier of biological information of potential value for mode of action risk assessment purposes by providing mechanistic insights into how and at which steps chemicals may disrupt essential metabolic pathways leading to eventual overt cellular toxicity. These types of data are highly useful in supporting mode of action–based risk assessments since major disturbances in metabolic pathways may ultimately lead to clinical disease states over time and the increased presence of metabolites in accessible matrices such as blood, urine, exhaled breath, and saliva or biopsy materials have proven to be highly useful as early biomarkers of ongoing cell injury processes in a number of situations.

Chemical-induced alterations in the heme biosynthetic pathway have proven useful for risk assessment purposes as a class of metabolomic biomarker for a number of important organic and inorganic toxic agents (Please see Silbergeld and Fowler, 1987 for a review). Arsenic in drinking water is an excellent example as discussed later.

Measurement of urinary porphyrins in persons exposed to arsenic in drinking water in Mexico has provided useful correlative health information linking exposure to arsenic to an adverse health effect for risk assessment purposes (Garcia-Vargas et al., 1994; Hernandez-Zavala et al., 1999). Data from these studies showed a strong correlation between concentrations if arsenic in drinking water and a porphyrinuria pattern similar to that previously reported in rodents exposed to arsenic in drinking water for a prolonged period of time (Woods and Fowler, 1978). These porphyrin patterns in rodents were linked to other cellular manifestations of toxicity in liver (Fowler and Woods, 1979; Fowler et al., 1979; Whittaker et al., 2011) providing useful correlative information on mechanisms of toxicity (eg, mode of action).

In addition, unique urinary porphyrin excretion patterns have also been identified in rats exposed to mixtures of lead, cadmium and arsenic in food (Fowler and Mahaffey, 1978; Mahaffey et al., 1981) and drinking water (Whittaker et al., 2011) at various dose levels under prolonged exposure conditions. Different porphyrin excretion patterns (Fig. 4.2) have been observed in hamsters dosed with different components of the binary semiconductor compound Indium arsenide suggesting that differential disturbances in heme pathway may be utilized as biomarkers for mixtures of other chemical agents (Fowler et al., 2005). Since the heme biosynthetic pathway is highly

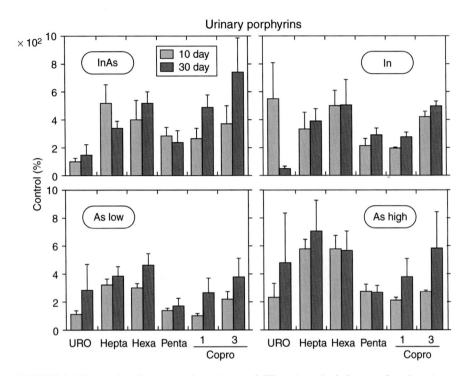

FIGURE 4.2 Composite urinary excretion patterns of different porphyrin isomers from hamsters exposed to InAs, As low (0.3 mg/kg), As high (3.0 mg/kg), or In at 10 or 30 days posttreatment. *Data from Conner et al. (1995).*

conserved across species (Panek and O'Brian, 2002), there is hence good reason to anticipate that the similar disturbances of this pathway elucidated in experimental systems could also be used for risk assessment purposes in humans under similar chemical mixture exposure conditions. Risk assessment for chemical mixtures is an important and difficult problem area for public health risk assessors and inclusion of molecular biomarker information such as that described earlier may provide a useful and needed information to address this problem area.

4.3 Microfluidics/Nanoproteomics

Going forward, the development of improved analytical methods using microfluidics capable of handling very small samples such as tissue biopsies or biological fluids has greatly increased the possibilities of conducting direct measurements for proteins of metabolites in samples of interest. This is particularly true for mass spectroscopy measurements using laser capture microdissection mass spectrometry.

5   CURRENT APPLICATIONS OF OMIC BIOMARKERS FOR RISK ASSESSMENT PURPOSES

Based upon the previous brief review of how genomic, proteomic, and metabolomic biomarkers in combination with computational toxicology approaches have been applied for risk assessment purposes in variety of chemical agents, it seems clear that these tools have already demonstrated their value in a number of important aspects of risk assessment practice. Specifically, these approaches have proven very effective in:

1. Delineating sensitive subpopulations under low-dose environmental exposure conditions to cadmium (Ruiz et al., 2010), lead (Scinicariello et al., 2010), and arsenic (Garcia-Vargas et al., 1994)
2. Providing clear evidence of early biological responses for these toxic elements on an individual or mixture basis which are in concert with tissue analytical data (Mahaffey et al., 1981; Whittaker et al., 2011).
3. Generating data which provide a basic scientific foundation for mode of action (MOA)–based risk assessment approaches (Demchuk et al., 2011)
4. Suggesting further avenues of hypothesis-generating molecular research of potential value in human epidemiological studies

5.1   Next Steps Forward for Incorporating Omics Approaches into Risk Assessment Practice

It should be clear from the previous examples that genomic, proteomic, and metabolomic/metabonomic biomarkers can each contribute useful mechanism-based risk assessment information both with regard to subpopulations at special risk but also providing correlative biological information with results of analytical chemistry exposure information such that more direct linkage may be made between a chemical exposure and adverse biological effect. In order for this goal to be achieved, collaborative studies between toxicologists, epidemiologists, analytical chemists, and risk assessors are needed. These types of multidisciplinary studies clearly fall into the translational/personalized biomedicine which is now being promulgated by a number of Federal agencies including the NIH (Collins, 2011; Collins and Varmus, 2015) and FDA (Hamburg and Collins, 2010) in order to obtain more precise risk assessment answers to a number of pressing public health issues involving toxic chemical and infectious agents.

References

Akutekwe, A., Seker, H., 2014. A hybrid dynamic Bayesian network approach for modelling temporal associations of gene expressions for hypertension diagnosis. Conf. Proc. IEEE Eng. Med. Biol. Soc. 2014, 804–807.

Alonso, A., Marsal, S., Julia, A., 2015. Analytical methods in untargeted metabolomics: state of the art in 2015. Front Bioeng. Biotechnol. 3, 23.

Bachmann, N., Spengler, S., Binder, G., Eggermann, T., 2010. MBD3 mutations are not responsible for ICR1 hypomethylatioj. Med in Silver–Russell syndrome. Eur. J. Med. Genetics 53 (1), 23–24.

Bakulski, K.M., Lee, H., Feinberg, J.I., Wells, E.M., Brown, S., Herbstman, J.B., et al., 2015. Prenatal mercury concentration is associated with changes in DNA methylation at TCEANC2 in newborns. Int. J. Epidemiol. 44, 1249–1262.

Bergdahl, I.A., Grubb, A., Schutz, A., Desnick, R.J., Wetmur, J.G., Sassa, S., Skerfving, S., 1997. Lead binding to delta-aminolevulinic acid dehydratase (ALAD) in human erythrocytes. Pharmacol. Toxicol. 81 (4), 153–158.

Bertrand, D., Chng, K.R., Sherbaf, F.G., Kiesel, A., Chia, B.K., Sia, Y.Y., et al., 2015. Patient-specific driver gene prediction and risk assessment through integrated network analysis of cancer omics profiles. Nucleic Acids Res. 43 (7), e44.

Bhattacharjee, P., Chatterjee, D., Singh, K.K., Giri, A.K., 2013. Systems biology approaches to evaluate arsenic toxicity and carcinogenicity: an overview. Int. J. Hyg. Environ. Health 216 (5), 574–586.

Biggar, K.K., Li, S.S., 2015. Non-histone protein methylation as a regulator of cellular signalling and function. Nat. Rev. Mol. Cell Biol. 16 (1), 5–17.

Boso, V., Herrero, M.J., Santaballa, A., Palomar, L., Megias, J.E., de la Cueva, H., et al., 2014. SNPs and taxane toxicity in breast cancer patients. Pharmacogenomics 15 (15), 1845–1858.

Boyd, A.S., Neldner, K.H., Naylor, M., 1989. 2, 3, 7, 8-Tetrachlorodibenzo-*p*-dioxin-induced porphyria cutanea tarda among pediatric patients. J. Am. Acad. Dermatol. 21 (6), 1320–1321.

Braundmeier, A.G., Lenz, K.M., Inman, K.S., Chia, N., Jeraldo, P., Walther-Antonio, M.R., et al., 2015. Individualized medicine and the microbiome in reproductive tract. Front Physiol. 6, 97.

Burke, A.M., Kandur, W., Novitsky, E.J., Kaake, R.M., Yu, C., Kao, A., et al., 2015. Synthesis of two new enrichable and MS-cleavable cross-linkers to define protein-protein interactions by mass spectrometry. Org. Biomol. Chem. 13 (17), 5030–5037.

Byrd, A.L., Segre, J.A., 2015. Integrating host gene expression and the microbiome to explore disease pathogenesis. Genome Biol. 16 (1), 70.

Cantoni, L., Dal Fiume, D., Ruggieri, R., 1984. Decarboxylation of uroporphyrinogen I and III in 2, 3, 7, 8-tetrachlorodibenzo-*p*-dioxin induced porphyria in mice. Int. J. Biochem. 16 (5), 561–565.

Chang, M.H., Lindegren, M.L., Butler, M.A., Chanock, S.J., Dowling, N.F., Gallagher, M., et al., 2009. Prevalence in the United States of selected candidate gene variants: Third National Health and Nutrition Examination Survey, 1991–1994. Am. J. Epidemiol. 169 (1), 54–66.

Chang, M.H., Yesupriya, A., Ned, R.M., Mueller, P.W., Dowling, N.F., 2010. Genetic variants associated with fasting blood lipids in the U.S. population: Third National Health and Nutrition Examination Survey. BMC Med. Genet. 11, 62.

Chappell, G., Kobets, T., O'Brien, B., Tretyakova, N., Sangaraju, D., Kosyk, O., et al., 2014. Epigenetic events determine tissue-specific toxicity of inhalational exposure to the genotoxic chemical 1,3-butadiene in male C57BL/6J mice. Toxicol. Sci. 142 (2), 375–384.

Collins, F.S., 2011. Reengineering translational science: the time is right. Sci. Transl. Med. 3 (90), 90cm17.

Collins, F.S., Varmus, H., 2015. A new initiative on precision medicine. N. Engl. J. Med. 372 (9), 793–795.

Combes, R.D., 2012. In silico methods for toxicity prediction. Adv. Exp. Med. Biol. 745, 96–116.

Conner, E.A., Yamauchi, H., Fowler, B.A., Akkerman, M., 1993. Biological indicators for monitoring exposure/toxicity from III-V semiconductors. J. Expo. Anal. Environ. Epidemiol. 3 (4), 431–440.

Conner, E.A., Yamauchi, H., Fowler, B.A., 1995. Alterations in the heme biosynthetic pathway from the III-V semiconductor metal, indium arsenide (InAs). Chem. Biol. Interact. 96 (3), 273–285.

Cote, I., Anastas, P.T., Birnbaum, L.S., Clark, R.M., Dix, D.J., Edwards, S.W., Preuss, P.W., 2012. Advancing the next generation of health risk assessment. Environ. Health Perspect. 120 (11), 1499–1502.

Demchuk, E., Ruiz, P., Chou, S., Fowler, B.A., 2011. SAR/QSAR methods in public health practice. Toxicol. Appl. Pharmacol. 254 (2), 192–197.

Dent, A.J., Beyersmann, D., Block, C., Hasnain, S.S., 1990. Two different zinc sites in bovine 5-aminolevulinate dehydratase distinguished by extended X-ray absorption fine structure. Biochemistry 29 (34), 7822–7828.

Donovan, M.J., Cordon-Cardo, C., 2014. Genomic analysis in active surveillance: predicting high-risk disease using tissue biomarkers. Curr. Opin. Urol. 24 (3), 303–310.

Dunagin, W.G., 1984. Cutaneous signs of systemic toxicity due to dioxins and related chemicals. J. Am. Acad. Dermatol. 10 (4), 688–700.

Ellis, P., Fowler, P., Booth, E., Kidd, D., Howe, J., Doherty, A., Scott, A., 2014. Where will genetic toxicology testing be in 30 years' time? Summary report of the 25th Industrial Genotoxicity Group Meeting, Royal Society of Medicine, London, November 9, 2011. Mutagenesis 29 (1), 73–77.

Fowler, B.A., Mahaffey, K.R., 1978. Interactions among lead, cadmium, and arsenic in relation to porphyrin excretion patterns. Environ. Health Perspect. 25, 87–90.

Fowler, B.A., Woods, J.S., 1979. The effects of prolonged oral arsenate exposure on liver mitochondria of mice: morphometric and biochemical studies. Toxicol. Appl. Pharmacol. 50 (2), 177–187.

Fowler, B.A., Woods, J.S., Schiller, C.M., 1979. Studies of hepatic mitochondrial structure and function: morphometric and biochemical evaluation of in vivo perturbation by arsenate. Lab. Invest. 41 (4), 313–320.

Fowler, B.A., Conner, E.A., Yamauchi, H., 2005. Metabolomic and proteomic biomarkers for III-V semiconductors: chemical-specific porphyrinurias and proteinurias. Toxicol. Appl. Pharmacol. 206 (2), 121–130.

Fowler, B.A., Conner, E.A., Yamauchi, H., 2008. Proteomic and metabolomic biomarkers for III-V semiconductors: and prospects for application to nano-materials. Toxicol. Appl. Pharmacol. 233 (1), 110–115.

Garcia-Vargas, G.G., Del Razo, L.M., Cebrian, M.E., Albores, A., Ostrosky-Wegman, P., Montero, R., et al., 1994. Altered urinary porphyrin excretion in a human population chronically exposed to arsenic in Mexico. Hum. Exp. Toxicol. 13 (12), 839–847.

Gim, J.A., Ha, H.S., Ahn, K., Kim, D.S., Kim, H.S., 2014. Genome-wide identification and classification of MicroRNAs derived from repetitive elements. Genomics Inform. 12 (4), 261–267.

Goering, P.L., Fowler, B.A., 1987a. Kidney zinc-thionein regulation of delta-aminolevulinic acid dehydratase inhibition by lead. Arch. Biochem. Biophys. 253 (1), 48–55.

Goering, P.L., Fowler, B.A., 1987b. Metal constitution of metallothionein influences inhibition of delta-aminolaevulinic acid dehydratase (porphobilinogen synthase) by lead. Biochem. J. 245 (2), 339–345.

Goodman, J.E., Boyce, C.P., Pizzurro, D.M., Rhomberg, L.R., 2014. Strengthening the foundation of next generation risk assessment. Regul. Toxicol. Pharmacol. 68 (1), 160–170.

Graziani, R., Guindani, M., Thall, P.F., 2015. Bayesian nonparametric estimation of targeted agent effects on biomarker change to predict clinical outcome. Biometrics 71 (1), 188–197.

Hack, C.E., Haber, L.T., Maier, A., Shulte, P., Fowler, B., Lotz, W.G., Savage, Jr., R.E., 2010. A Bayesian network model for biomarker-based dose response. Risk Anal. 30 (7), 1037–1051.

Hamburg, M.A., Collins, F.S., 2010. The path to personalized medicine. N. Engl. J. Med. 363 (4), 301–304.

Hernandez-Zavala, A., Del Razo, L.M., Garcia-Vargas, G.G., Aguilar, C., Borja, V.H., Albores, A., Cebrian, M.E., 1999. Altered activity of heme biosynthesis pathway enzymes in individuals chronically exposed to arsenic in Mexico. Arch. Toxicol. 73 (2), 90–95.

Hughes, D.A., Kircher, M., He, Z., Guo, S., Fairbrother, G.L., Moreno, C.S., et al., 2015. Evaluating intra- and inter-individual variation in the human placental transcriptome. Genome Biol. 16 (1), 54.

Ilyas, M., Kim, J.S., Cooper, J., Shin, Y.A., Kim, H.M., Cho, Y.S., et al., 2015. Whole genome sequencing of an ethnic Pathan (Pakhtun) from the north-west of Pakistan. BMC Genomics 16, 172.

Isokpehi, R.D., Udensi, U.K., Anyanwu, M.N., Mbah, A.N., Johnson, M.O., Edusei, K., et al., 2012. Knowledge building insights on biomarkers of arsenic toxicity to keratinocytes and melanocytes. Biomark Insights 7, 127–141.

Kamitsuji, S., Matsuda, T., Nishimura, K., Endo, S., Wada, C., Watanabe, K., et al., 2015. Japan PGx Data Science Consortium Database: SNPs and HLA genotype data from 2994 Japanese healthy individuals for pharmacogenomics studies. J. Hum. Genet. 60, 319–326.

Kelly, R.S., Vineis, P., 2014. Biomarkers of susceptibility to chemical carcinogens: the example of non-Hodgkin lymphomas. Br. Med. Bull. 111 (1), 89–100.

Khoury, M.J., Bowen, S., Bradley, L.A., Coates, R., Dowling, N.F., Gwinn, M., et al., 2009. A decade of public health genomics in the United States: centers for disease control and prevention 1997–2007. Public Health Genomics 12 (1), 20–29.

Kleensang, A., Maertens, A., Rosenberg, M., Fitzpatrick, S., Lamb, J., Auerbach, S., et al., 2014. t4 Workshop report: pathways of toxicity. ALTEX 31 (1), 53–61.

Kloosterman, W.P., Francioli, L.C., Hormozdiari, F., Marschall, T., Hehir-Kwa, J.Y., Abdellaoui, A., et al., 2015. Characteristics of de novo structural changes in the human genome. Genome Res. 25, 792–801.

Kouzarides, T., 2007. Chromatin modifications and their function. Cell 128 (4), 693–705.

Krewski, D., Westphal, M., Andersen, M.E., Paoli, G.M., Chiu, W.A., Al-Zoughool, M., et al., 2014. A framework for the next generation of risk science. Environ. Health Perspect. 122 (8), 796–805.

Krieg, Jr., E.F., Butler, M.A., Chang, M.H., Liu, T., Yesupriya, A., Lindegren, M.L., Dowling, N., 2009. Lead and cognitive function in ALAD genotypes in the third National Health and Nutrition Examination Survey. Neurotoxicol. Teratol. 31 (6), 364–371.

Krieg, Jr., E.F., Butler, M.A., Chang, M.H., Liu, T., Yesupriya, A., Dowling, N., Lindegren, M.L., 2010. Lead and cognitive function in VDR genotypes in the third National Health and Nutrition Examination Survey. Neurotoxicol. Teratol. 32 (2), 262–272.

Lee, H.-E., Kim, E.-A., Park, J., Kang, S.-K., 2011. Cancer mortality and incidence in Korean semiconductor workers. Saf. Health Work 2 (2), 135–147.

Liao, W., Stahle, M., Franke, A., Zhang, X., Liu, J., Marusic, Z., et al., 2015. Histomorphologic spectrum of Bap1 negative melanocytic neoplasms in a family with Bap1-associated cancer susceptibility syndrome. Nat. Commun. 42, 406–412.

Lin, C.-C., Wang, J.-D., Hsieh, G.-Y., Chang, Y.-Y., Chen, P.-C., 2008. Health risk in the offspring of female semiconductor workers. Occup. Med. 58 (6), 388–392.

Mahaffey, K.R., Capar, S.G., Gladen, B.C., Fowler, B.A., 1981. Concurrent exposure to lead, cadmium, and arsenic. Effects on toxicity and tissue metal concentrations in the rat. J. Lab. Clin. Med. 98 (4), 463–481.

McConnell, E.R., Bell, S.M., Cote, I., Wang, R.L., Perkins, E.J., Garcia-Reyero, N., et al., 2014. Systematic omics analysis review (SOAR) tool to support risk assessment. PLoS One 9 (12), e110379.

McEwen, B.S., 2015. Biomarkers for assessing population and individual health and disease related to stress and adaptation. Metabolism 64 (3 Suppl. 1), S2–S10.

Meek, M.E., Boobis, A., Cote, I., Dellarco, V., Fotakis, G., Munn, S., et al., 2014. New developments in the evolution and application of the WHO/IPCS framework on mode of action/species concordance analysis. J. Appl. Toxicol. 34 (1), 1–18.

Mischak, H., Critselis, E., Hanash, S., Gallagher, W.M., Vlahou, A., Ioannidis, J.P., 2015. Epidemiologic design and analysis for proteomic studies: a primer on -omic technologies. Am. J. Epidemiol. 181 (9), 635–647.

NAS/NRC, 1993. Measuring Lead Exposure in Infants, Children and Other Sensitive Populations. National Academies Press, Washington, DC, pp. 337.

Negi, S., Pandey, S., Srinivasan, S.M., Mohammed, A., Guda, C., 2015. LocSigDB: a database of protein localization signals. Database 2015, bav003.

Neslund-Dudas, C., Levin, A.M., Rundle, A., Beebe-Dimmer, J., Bock, C.H., Nock, N.L., et al., 2014. Case-only gene-environment interaction between ALAD tagSNPs and occupational lead exposure in prostate cancer. Prostate 74 (6), 637–646.

Ng, E., Salihovic, S., Monica Lind, P., Mahajan, A., Syvanen, A.C., Axelsson, T., et al., 2015. Genome-wide association study of plasma levels of polychlorinated biphenyls disclose an association with the CYP2B6 gene in a population-based sample. Environ. Res. 140, 95–101.

Ngalame, N.N., Micciche, A.F., Feil, M.E., States, J.C., 2013. Delayed temporal increase of hepatic Hsp70 in ApoE knockout mice after prenatal arsenic exposure. Toxicol. Sci. 131 (1), 225–233.

Pan, W.C., Kile, M.L., Seow, W.J., Lin, X., Quamruzzaman, Q., Rahman, M., et al., 2013. Genetic susceptible locus in NOTCH2 interacts with arsenic in drinking water on risk of type 2 diabetes. PLoS One 8 (8), e70792.

Panagiotou, G., Taboureau, O., 2012. The impact of network biology in pharmacology and toxicology. SAR QSAR Environ. Res. 23 (3–4), 221–235.

Panek, H., O'Brian, M.R., 2002. A whole genome view of prokaryotic haem biosynthesis. Microbiology 148 (Pt. 8), 2273–2282.

Papageorgiou, A., Rapley, J., Mesirov, J.P., Tamayo, P., Avruch, J., 2015. A genome-wide siRNA screen in mammalian cells for regulators of S6 phosphorylation. PLoS One 10 (3), e0116096.

Patel, C.J., Cullen, M.R., 2012. Genetic variability in molecular responses to chemical exposure. EXS 101, 437–457.

Pierce, B.L., Kibriya, M.G., Tong, L., Jasmine, F., Argos, M., Roy, S., et al., 2012. Genome-wide association study identifies chromosome 10q24.32 variants associated with arsenic metabolism and toxicity phenotypes in Bangladesh. PLoS Genet. 8 (2), e1002522.

Ramirez-Santana, M., Zuniga, L., Corral, S., Sandoval, R., Scheepers, P.T., Van der Velden, K., et al., 2015. Assessing biomarkers and neuropsychological outcomes in rural populations exposed to organophosphate pesticides in Chile—study design and protocol. BMC Public Health 15, 116.

Ruiz, P., Mumtaz, M., Osterloh, J., Fisher, J., Fowler, B.A., 2010. Interpreting NHANES biomonitoring data, cadmium. Toxicol. Lett. 198 (1), 44–48.

Schrattenholz, A., Soskic, V., Schopf, R., Poznanovic, S., Klemm-Manns, M., Groebe, K., 2012. Protein biomarkers for in vitro testing of toxicology. Mutat. Res. 746 (2), 113–123.

Scinicariello, F., Portier, C., 2015. A simple procedure for estimating pseudo risk ratios from exposure to non-carcinogenic chemical mixtures. Arch. Toxicol. 90, 513–523.

Scinicariello, F., Murray, H.E., Smith, L., Wilbur, S., Fowler, B.A., 2005. Genetic factors that might lead to different responses in individuals exposed to perchlorate. Environ. Health Perspect. 113 (11), 1479–1484.

Scinicariello, F., Yesupriya, A., Chang, M.H., Fowler, B.A., 2010. Modification by ALAD of the association between blood lead and blood pressure in the U.S. population: results from the Third National Health and Nutrition Examination Survey. Environ. Health Perspect. 118 (2), 259–264.

Silbergeld, E.K., Fowler, B.A. (Eds.), 1987. Mechanisms of chemical—induced porphyrinopathies. Ann. N Y Acad. Sci. 514, 352.

Simon, T.W., Simons, Jr., S.S., Preston, R.J., Boobis, A.R., Cohen, S.M., Doerrer, N.G., et al., 2014. The use of mode of action information in risk assessment: quantitative key events/dose-response framework for modeling the dose-response for key events. Crit. Rev. Toxicol. 44 (Suppl. 3), 17–43.

Smedley, D., Haider, S., Durinck, S., Pandini, L., Provero, P., Allen, J., et al., 2015. The BioMart community portal: an innovative alternative to large, centralized data repositories. Nucleic Acids Res. 43, W589–W598.

Sobin, C., Parisi, N., Schaub, T., Gutierrez, M., Ortega, A.X., 2011. delta-Aminolevulinic acid dehydratase single nucleotide polymorphism 2 and peptide transporter 2*2 haplotype may differentially mediate lead exposure in male children. Arch. Environ. Contam. Toxicol. 61 (3), 521–529.

Tseng, Y.T., Li, W., Chen, C.H., Zhang, S., Chen, J.J., Zhou, X., Liu, C.C., 2015. IIIDB: a database for isoform-isoform interactions and isoform network modules. BMC Genomics 16 (Suppl. 2), S10.

Tzou, W.S., Chu, Y., Lin, T.Y., Hu, C.H., Pai, T.W., Liu, H.F., et al., 2014. Molecular evolution of multiple-level control of heme biosynthesis pathway in animal kingdom. PLoS One 9 (1), e86718.

van Bemmel, D.M., Li, Y., McLean, J., Chang, M.H., Dowling, N.F., Graubard, B., Rajaraman, P., 2011. Blood lead levels, ALAD gene polymorphisms, and mortality. Epidemiology 22 (2), 273–278.

Welch, M.A., Kock, K., Urban, T.J., Brouwer, K.L., Swaan, P.W., 2015. Toward predicting drug-induced liver injury: parallel computational approaches to identify multidrug resistance protein 4 and bile salt export pump inhibitors. Drug Metab. Dispos. 43 (5), 725–734.

Wetmur, J.G., Kaya, A.H., Plewinska, M., Desnick, R.J., 1991. Molecular characterization of the human delta-aminolevulinate dehydratase 2 (ALAD2) allele: implications for molecular screening of individuals for genetic susceptibility to lead poisoning. Am. J. Hum. Genet. 49 (4), 757–763.

Whittaker, M.H., Wang, G., Chen, X.Q., Lipsky, M., Smith, D., Gwiazda, R., Fowler, B.A., 2011. Exposure to Pb, Cd, and As mixtures potentiates the production of oxidative stress precursors: 30-day, 90-day, and 180-day drinking water studies in rats. Toxicol. Appl. Pharmacol. 254 (2), 154–166.

Williams, S.B., Ye, Y., Huang, M., Chang, D.W., Kamat, A.M., Pu, X., et al., 2015. Mitochondrial DNA content as risk factor for bladder cancer and its association with mitochondrial DNA polymorphisms. Cancer Prev. Res. 8, 607–613.

Woods, J.S., Fowler, B.A., 1978. Altered regulation of mammalian hepatic heme biosynthesis and urinary porphyrin excretion during prolonged exposure to sodium arsenate. Toxicol. Appl. Pharmacol. 43 (2), 361–371.

Xia, J., Sinelnikov, I.V., Han, B., Wishart, D.S., 2015. MetaboAnalyst 3.0-making metabolomics more meaningful. Nucleic Acids Res. 43, W251–W257.

Yin, X., Low, H.Q., Wang, L., Li, Y., Ellinghaus, E., 2015. Genome-wide meta-analysis identifies multiple novel associations and ethnic heterogeneity of psoriasis susceptibility 6, 6916.

Zhang, H., Kawase-Koga, Y., Sun, T., 2015a. Protein expression profiles characterize distinct features of mouse cerebral cortices at different developmental stages. PLoS One 10 (4), e0125608.

Zhang, C., Yan, D., Wang, S., Xu, C., Du, W., Ning, T., Chen, Z., 2015b. Genetic polymorphisms of NAMPT related with susceptibility to esophageal squamous cell carcinoma. BMC Gastroenterol 15 (1), 49.

Zhao, Y.Y., Tang, D.D., Chen, H., Mao, J.R., Bai, X., Cheng, X.H., Xiao, X.Y., 2015. Urinary metabolomics and biomarkers of aristolochic acid nephrotoxicity by UPLC-QTOF/HDMS. Bioanalysis 7 (6), 685–700.

Zou, D., Ma, L., Yu, J., Zhang, Z., 2015. Biological databases for human research. Genomics Proteomics Bioinformatics 13 (1), 55–63.

CHAPTER 5

Validation of Biological Markers for Epidemiological Studies

1 INTRODUCTION

As already noted in this book, technical advances in analytical chemistry, molecular biology, and computational biology have provided new insights into chemical exposure–induced perturbations of normal cellular biology and ever lower dose levels. In order to utilize such information for risk assessment purposes, it is imperative to understand the biological consequences of these changes. Such information is essential for mechanism or mode of action (MOA)-based risk assessments (Cote et al., 2012; Fowler, 2012; Meek et al., 2014). In general terms, the MOA approach is comprised of understanding linkages between drug or chemical exposures and adverse outcome pathways (AOPs) (Becker et al., 2015; Burden et al., 2015; Phillips et al., 2015; Vinken, 2015; Yauk et al., 2015), which will lead to clinical disease or cancer and using this basic scientific information to improve the predictive precision of risk assessments. In order to achieve this goal, it is essential to have data from correlative validation studies for potential molecular biomarkers at the cell, organ, tissue organism, and population levels of biological organization. The overarching goal which must be addressed by these correlative validation studies concerns assessing the biological significance of a change in a putative biomarker and hence its prognostic utility. There are a number of ways to achieve this goal at the various levels of biological organization and these are discussed later. The approaches used will depend on the scientific or public health issues of interest. It is, of course, possible that several validation methods may be utilized in concert and integrated with each other through the use of computational toxicology methods. This chapter will focus on a number of biology-based validation methods for biomarkers and refer the reader to other references and chapters in this book series for a discussion of integration methods (eg, computational toxicology). Ideally, one would ultimately like to have an integrated picture across the various levels of biological organization in order to reach a more comprehensive and robust risk assessment conclusion.

CONTENTS

Molecular Biological Markers for Toxicology and Risk Assessment. http://dx.doi.org/10.1016/B978-0-12-809589-8.00005-6

It should be noted that several interrelated AOPs may be simultaneously operating as a function of chemical/drug dose, duration of exposure, and other modifying factors such as gender, age, genetics, and nutritional status. These sorts of concomitant effectors may complicate accurate interpretation of observed health outcomes with regard to the primary AOP. The evolving tools of computational toxicology may prove of particular value in this regard, going forward in time.

2 MOLECULAR BIOMARKER VALIDATION THROUGH CORRELATION WITH OTHER BIOLOGICAL ENDPOINTS

2.1 Cellular Level

2.1.1 Morphological Methods

2.1.1.1 Histopathology

Classical histopathology techniques for identifying correlations with biochemical markers have been used for many years as prognostic indicators of various clinical diseases and cancer in humans and animals. Histopathology evaluations have been used successfully by the National Toxicology Program (NTP) to provide valuable interpretive information for rodent toxicity and long-term cancer studies (National Toxicology Program, 2004, 2011). This approach has also been used in a number of cancer epidemiology studies on human populations with known chemical exposure (Cogliano et al., 2011; Grosse et al., 2013; Straif et al., 2009).

2.1.1.2 Ultrastructural Morphometry

Ultrastructural morphometry is a technique for quantifying organelles within intact cells of target organs such as the kidney (Fowler, 1983). This basic cell biology approach coupled with correlative biochemical measurements of isolated organelles showing physical changes (Fowler, 1980) has proven useful in providing quantitative intracellular data linking chemical–induced alterations in target organelles with changes in molecular biomarkers (Fowler, 1983; Hegstad et al., 1994; Stacchiotti et al., 2002). Such information provides a basic scientific foundation for MOA since many of the sources of circulating or excreted biomarker endpoints ultimately originate from target organelles (eg, mitochondria) in target organs such as the liver and kidney (Fowler et al., 1975, 1979, 1983; Fowler and Nordberg, 1978; Fowler and Woods, 1977, 1979; Woods and Fowler, 1986). Such data are clearly useful for interpreting biomarker responses since they link these changes with ultrastructural pathology in organelles in target cell populations associated with the biomarker response of interest. This information is also of value in generating mechanism-based hypotheses for advancing the fields of toxicology and risk assessment. For example, morphometric analyses which demonstrated marked

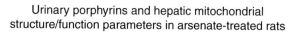

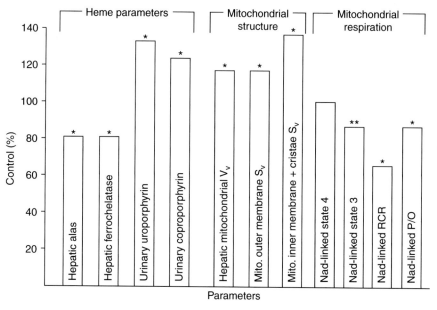

Urinary porphyrins and hepatic mitochondrial
structure/function parameters in arsenate-treated rats

*-Differs from control $P < .05$
**-Differs from control $P < .06$

FIGURE 5.1 Composite figure of ultrastructural morphometric data from rats exposed to arsenate in drinking water for 6 weeks showing statistical swelling and proliferation of the mitochondrial inner membrane and cristae. *From Fowler et al. (1987).*

proliferation of the liver mitochondrial inner membranes (Fig. 5.1) in concert with high amplitude swelling and development of disturbances of other biochemical functions localized in the mitochondrial inner membrane. These structural perturbations were associated with porphyrinuria in rats and mice (Fowler et al., 1979; Woods and Fowler, 1978) exposed to arsenate in drinking water for prolonged time periods. The value of these data rests with establishment of linkages between physical changes in the structure of an essential organelle, biochemical alterations in a number of biochemical functions localized in those intraorganelle compartments and development of a metabolomic biomarker change (eg, arsenic-specific porphyrinuria pattern). These findings subsequently lead to identification of similar porphyrinuria patterns in humans exposed to arsenic contaminated drinking water (Garcia-Vargas et al., 1994). These basic scientific data hence provide a MOA interpretation for the observed arsenic-induced porphyrinuria pattern in humans thus increasing the value of this metabolomic biomarker for risk assessment purposes.

2.1.2 Biochemical Methods

2.1.2.1   Serum/Urinary Enzyme Activities or Protein Patterns

Measurement of changes in the activities of enzymes in blood, serum, or urine or specific proteins in these matrices has been used for many years as the basis for clinical decision making in the diagnosis and treatment of many diseases and cancers (Bagrodia et al., 2014; Han et al., 2015; Panotopoulos et al., 2015; Tan et al., 2015; Tikoo et al., 2015; Wu et al., 2015). In addition, the ability to distinguish tissue- or organ-specific isozymes has allowed clinicians to distinguish tissue sources of measured alterations in the activities of enzymes on a very targeted basis (Bagrodia et al., 2014; Barfeld et al., 2014; Han et al., 2015; Panotopoulos et al., 2015; Tan et al., 2015; Tikoo et al., 2015; Wu et al., 2015). One study (Scinicariello et al., 2010) used differences in alleles for the heme pathway enzyme ALAD (ALAD1, ALAD2, ALAD1/2), which is the main carrier for lead in blood, to provide useful information on the relationships between blood lead values, ALAD genotype, and elevated risk of lead-induced hypertension associated with the ALAD 2 allele. Similar effects have been observed for kidney toxicity in relation to plasma lead concentrations modified by ALAD 2 polymorphism (Tian et al., 2013). Similarly, genetic difference alleles of enzymes such as glutathione S-transferase (GSTs) have been shown (Cong et al., 2014; Deng et al., 2015; Economopoulos and Sergentanis, 2010; Fan et al., 2013; Fletcher et al., 2015; Jaramillo-Rangel et al., 2015; Kassogue et al., 2015; Kruger et al., 2015; Song et al., 2012) to produce variable differences in susceptibility to develop different types of cancer depending on a number of factors such as racial group. Measurement of specific proteins in urine such as albumin (Abe et al., 2015; Grams et al., 2015; Joshi and Viljoen, 2015; Kataria et al., 2015; Yoon et al., 2015), beta-2 microglobulin (Shin et al., 2014), retinol binding protein (Pallet et al., 2014), and more recently kidney injury molecule-1 [KIM-1 (Wunnapuk et al., 2013)] or proteinuria patterns visualized by silver staining of 2-D gels (Candiano et al., 2010; Chen et al., 2011; Klawitter et al., 2010; Varghese et al., 2010) have been used to assess renal damage from nephrotoxic agents on an individual or chemical mixture basis (Fowler et al., 2005, 2008).

2.2   Correlations With Chemical Measurements of Pharmaceutical or Toxic Agents

Correlations of analytical chemistry measurements for pharmaceutical or toxic agents with expected clinical outcomes have been extremely valuable at both a clinical and a public health level in tracking, treating, and preventing adverse health outcomes. For example, these types of data from the ongoing NHANES studies (CDC, 2005) have had fundamental impacts on the management and regulation of chemical exposures to major toxic agents such as lead, cadmium, mercury, and arsenic. This information has helped to guide a number of public health for these elements and the efficacy of implemented prevention strategies. The coupling of these baseline data, which are representative of the general US population, with sophisticated molecular biomarker information capable of identifying sensitive subpopulations at special risk for toxicity should greatly increase the efficacy of future risk assessments using these data.

2.3 Correlation With Tissue Concentrations and Intracellular Binding Patterns of Chemicals

In addition to accurately knowing the total concentration of a given agent in the blood, tissue, or urine, it is important to understand how it is bound in those matrices and hence its bioavailability to biomarker endpoints. One example is that of blood ALAD, which is the major binding protein for lead in blood (Bergdahl et al., 1997; Tian et al., 2013). The activity of this enzyme is also extremely sensitive to lead inhibition. It is hence an excellent biomarker for biomonitoring the biological activity of lead. It should be noted that the ability of lead to inhibit the activity of ALAD may be altered by other metals such as zinc since it is a zinc-dependent/zinc-activated enzyme (Goering and Fowler, 1987a,b,c).

Another important aspect of biomarker validation is correlation of the biomarker response with tissue or target cellular concentration of the administered chemical or its metabolites as a function of dose and time. The tissue binding patterns of toxic agents are dynamic and changes in binding patterns as a function of dose and time may have a profound impact on an ultimate health outcome. This is an important consideration for interpretation of a putative biomarker since it may mean that a characteristic biomarker response may appear shortly after exposure and then abate as a result of induced metabolism or increased binding due to induction of intracellular binding entities such as glutathione (Agrawal et al., 2014; Feng and He, 2013; Kim et al., 2015; Wu et al., 2014) or metal-binding proteins such as the metallothioneins (Fowler et al., 1987; Goering and Fowler, 1987a,b,c), low molecular weight lead binding proteins (Goering and Fowler, 1985, 1987c; Goering et al., 1986; Quintanilla-Vega et al., 1995; Smith et al., 1998), intranuclear inclusion bodies (Fowler et al., 1980), or mineral concretions (Carmichael and Fowler, 1981). Conversely, the biomarker response may appear or reappear after prolonged exposures or elevated dose levels due to overcoming normal cellular defense mechanisms (Fowler, 2009; Fowler et al., 2005; Whittaker et al., 2011).

2.4 Impacts on Biomarker Responses in Mixture Exposure Conditions

In cases of mixture exposures, competition or displacement among mixture components for critical intracellular binding sites may disrupt normal or

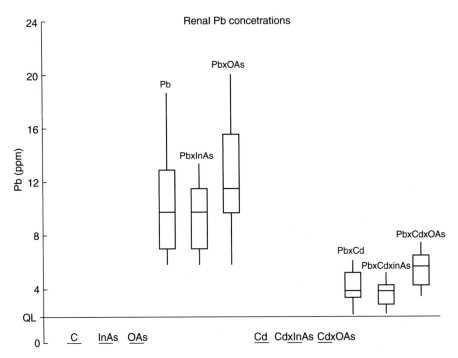

FIGURE 5.2 Composite diagram of kidney lead concentrations in rats with dietary exposures to lead, cadmium, and arsenic showing marked decreases in rat exposed to cadmium. *Data from Mahaffey et al. (1981).*

expected dose-response relationships and result in lower tissue concentrations of the mixture elements but enhanced biomarker responses due to increased intracellular bioavailability of reactive chemical species. One example is the common situation of concomitant exposures to lead, cadmium, and arsenic under occupational or environmental conditions. Experimental studies in rats under stressor dose levels (Fowler and Mahaffey, 1978; Mahaffey et al., 1981) or no observed effect level (NOEL) dose levels (Whittaker et al., 2011) showed marked reduction in tissue concentrations of lead in rats exposed to cadmium (Fig. 5.2). These changes were correlated with reduction of pathognomonic lead intranuclear inclusion bodies (Figs. 5.3 and 5.4) which contain high levels of lead (Fowler, 1980). These lead tissue concentration changes were associated with increased changes in heme biosynthetic pathway biomarker responses including elevated excretion of ALA and porphyrins in the urine (Fig. 5.5). The net result is reduced total tissue concentrations of lead but increased bioavailability of lead to sensitive biomarker systems such as the heme biosynthetic pathway. For these reasons, it is important to have basic scientific knowledge on both the mechanisms of the intracellular handling of the chemical species of interest as a function of dose and time in relation to the biomarker response

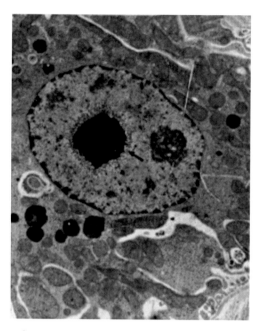

FIGURE 5.3 Electron micrograph of lead intranuclear inclusion body from a rat exposed to dietary lead (Fowler and Mahaffey, 1978).

Relative incidence of lead intranuclear inclusion bodies in rats fed diets containing lead, cadmium, inorganic, or organic arsenic

	Incidence[a]
Control	0/13
Pb	10/13
Cd	0/13
Inorg. As	0/13
Org As	0/13
Pb × Cd	0/13
Pb × Inorg. As	10/13
Pb × Org. As	11/13
Cd × Inorg. As	0/13
Cd × Org. As	0/13
Pb × Cd x Inorg. As	0/13
Pb × Cd x Org. As	2/13

[a]Number of rats with inclusions over number examined.

FIGURE 5.4 Summary of incidence of lead intranuclear inclusion body formation in rats exposed to dietary lead, cadmium, and arsenic for 10 weeks. *Data from Fowler and Mahaffey (1978).*

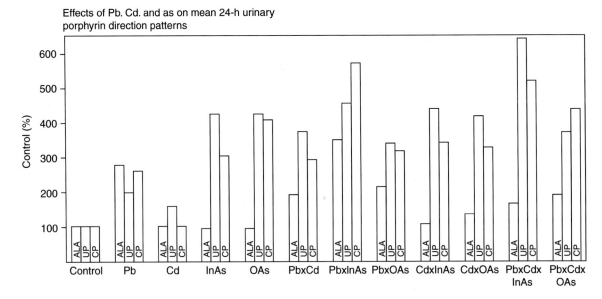

FIGURE 5.5 Porphyrinuria patterns from rats exposed to lead, cadmium, and arsenic in the diet for 10 weeks showing differences in mixture patterns and increased total porphyrins in rats, given the lead × cadmium dietary mixture (Fowler and Mahaffey, 1978).

of interest. This information is critical so that a correct interpretation of the molecular biomarker response may be ascertained and also used to help interpret the analytical data on tissue concentrations of the elements of interest.

2.5   Connectivity With Cell Death/Adverse Outcome Pathways

In recent years, a great deal of attention has been focused on the processes by which cells die. Necrosis is usually regarded as an overt set of processes by which groups of cells die in a short time period. This mode of cell death is typically observed in relation to acute events such as anoxia or acute high dose chemical/drug exposures which overwhelm cellular defense mechanisms and destroy large groups of cells or tissues leading to extensive organ damage or failure. On the other hand, apoptosis (Saikumar et al., 1999) or programmed cell death mediated by the caspase cascade is a more selective set of processes leading to cell death involving the dismantling of basic cellular machinery in only selected cells leading to their removal from the tissue. Apoptosis is a normal cellular process which is integral to tissue remodeling, growth and aging (Fowler, 2013a) and may be triggered and regulated by a number of mechanisms. This set of processes may also lead to initiation of carcinogenesis as a consequence of cell death and replacement. The scientific literature on the mechanisms of apoptosis-induced cell death is extremely large and expanding rapidly and the reader is referred to a number of authoritative reference works (Chen and Wang, 2002; Orrenius et al., 2003; Ott et al., 2007) on this subject area which include a discussion of a number of molecular biomarkers for these processes.

2.6 Molecular Biomarker Development via In Vitro or Alternative Animal Model Test Systems

In order for molecular biomarkers to be useful for human public health risk assessments, it is important that these responses be conserved across species. This means that a putative molecular biomarker response delineated in an in vitro, experimental mammalian model, or alternative animal test systems is also found in humans. Some degree of variation between species may occur, due to factors such as different epigenetic control systems, body temperature, and impacts of age or gender but the overall biological response is conserved. The net result will be that observed molecular biomarker responses may not be exactly identical with an in vivo analog from humans, but the responses are usually very similar and hence biologically applicable for providing MOA information of interpretive value for risk assessment purposes. Such mutually consistent interspecies or test system data are very valuable in providing confidence that the observed biomarker response is representative of a consistent and reliable biological response to a given chemical exposure and may hence be evaluated further as the basis of a molecular biomarker test system. For example, lead inhibits ALAD in both fish species (Conner and Fowler, 1994) and humans (NAS/NRC, 1993) but the rate of ALAD catalysis is lower in fish due to lower body temperature. The sensitivity of ALAD to lead inhibition is hence conserved across species. The utility of the ALAD as a sensitive biomarker for lead toxicity is hence also conserved across phyla. Since heme synthesis is essential for aerobic metabolism and life in most species, inhibition of the enzymes in the heme biosynthetic pathway and attendant increases in heme precursor intermediates, which are capable of catalyzing formation of reactive oxygen species, may hence be regarded as adverse outcome events. Chronic decreases in heme biosynthesis would ultimately compromise the ability of an organism to produce energy via aerobic metabolism and alter the functionality of heme-dependent monooxygenase systems which metabolize a host of endogenous and xenobiotic substances including carcinogens. The increased presence of heme precursors capable of contributing to oxidative stress would also be expected to contribute to decreased cell viability and initiation of carcinogenic processes (De Siervi et al., 2002; Ryter and Choi, 2013; Ryter and Tyrrell, 2000). A compromise of the ability of cells to produce heme could hence be regarded as an AOP due to inhibition of a pathway which is essential to life.

3 APPLICATION OF COMPUTATIONAL MODELING APPROACHES FOR EXTRAPOLATING FROM IN VITRO OR EXPERIMENTAL ANIMAL MODEL SYSTEMS FOR MOLECULAR BIOMARKER VALIDATIONS

There is a large body of mechanistic toxicology data from both in vitro and experimental animal model systems which could be used, in principle, to develop and validate MOA-based biomarkers for human risk assessment purposes. The advent of computational toxicology methods, which are discussed in another volume of this series (Fowler, 2013b) represents a promising set of tools for facilitating the translation of basic scientific toxicology information for risk assessment purposes. These tools have the virtue of speed, relatively low cost, and the ability to readily share information between investigators. All these factors, coupled with rapid online peer-review publishing will continue to greatly expedite the evolution of biomarker development. It is also worth noting that computational toxicology methods are themselves undergoing rapid expansion so that the pace and acceptance of molecular biomarkers as tools for modern risk assessment practice can be expected to continue to accelerate. The net result of these advances over the next decade should be even greater reliance on cell–human/animal–human MOA data for risk assessment decision making.

4 EASE OF APPLICATION FOR RISK ASSESSMENT PRACTICE

In order for a molecular biomarker to gain widespread acceptance and usage for risk assessment purposes, it must have relatively low cost, possess ease of measurement, be amenable to automated data handling, and have its relationships to mechanisms of cell injury and cell death. In practical terms, this means that molecular biomarkers which can be developed into ease of use kits, with relatively stable components, will be more likely to gain widespread acceptance and application for human field studies. This feature would be of particular value in remote areas of developing countries with limited access to refrigeration or consistent electrical power. For example, measurement of ALA in urine and erythrocyte zinc protoporphyrin would be examples of these types of tests (Whittaker et al., 2011). The overall point here is that ease of use, modest costs and stability of reagents are important considerations in determining the extent to which a putative biomarker may be used in real world situations. The generic policy issue of how to gain acceptance of biomarkers and other modern tools which increase precision in clinical medicine and public health practice has been recently reviewed by Kohane (2015).

5 CORRELATIONS AT THE POPULATION LEVEL AND POPULATION-BASED RISK ASSESSMENT STUDIES VIA NHANES

5.1 Data Mining

In recent years, a number of investigators (Tellez-Plaza et al., 2012a,b; Yeh et al., 2015; Zheng et al., 2014) have taken advantage of the multidisciplinary analytical/clinical chemistry data in the NHANES database to look for associations between exposures to chemicals or drugs and biological alterations in standard clinical chemistry tests. While many of the clinical chemistry tests have limited sensitivity and specificity, the value of these studies rests with a solid understanding of the relationships between changes in these parameters and clinical disease outcomes and the fact that the data are representative of a cross-section of the general US population. These data are hence reflective of different racial groups by gender across an age range from young children to the elderly. For these reasons, this information is of potentially great value for global risk assessments of the general US population. In addition, since a large number of blood and urine samples are archived, it is hence possible to reanalyze a number of samples with newer and more sophisticated molecular biomarker tests as these tests become available and validated. Another potentially great value of the NHANES data set is that since these studies have been conducted over a number of decades, it is potentially possible to look for trends in the general US population over time (NAS/NRC, 1993; Pirkle et al., 1994, 1998). A good example of this application would be tracking the steady decline in lead concentrations in blood over past decades (Pirkle et al., 1994, 1998) (NAS/NRC, 1993). These data not only show the impact of the removal of lead from gasoline but also that lead exposures persist in the general US population from other sources such as lead-based paint dust (Etchevers et al., 2015; Gulson et al., 2013, 2014; Laidlaw et al., 2014; Wilson et al., 2015), lead solder in domestic drinking water systems (Akers et al., 2015; Etchevers et al., 2015; Ngueta et al., 2015, 2014), and emissions from industrial sources such as coal-fired power plants, municipal incinerators, and lead smelters (NAS/NRC, 1993) both inside and outside the borders of the United States as a result of atmospheric transport and deposition of lead particulates (Gallon et al., 2011).

In addition, as noted earlier, the NHANES database has been used to identify subpopulations at special risk for toxicity from metals such as cadmium as function of gender and age (Ruiz et al., 2010) and lead as a function of genetic inheritance for alleles of enzymes such as ALAD (Scinicariello et al., 2010) which is both a major carrier protein for lead in blood and a sensitive biomarker for lead inhibition.

Other NHANES data mining studies have focused on geographic regional differences in population exposures to mercury (Mahaffey et al., 2009) and persistent organic pollutants (POPs) to identify risks for subpopulations at special risk for adverse health outcomes such as insulin resistance and development of Type II diabetes (Lee et al., 2006, 2007; Taylor et al., 2013).

Taken together, the previous collection of population-based data sets provide valuable information on subpopulations at special risk for specific chemical exposures and health outcomes as function of age, gender, and regional geographic location. These subpopulations are hence also prime groups for other validation studies of new and evolving molecular biomarkers capable of further elucidating underlying mechanisms and supporting MOA risk assessments as discussed later since this data set also includes information on genetic alleles for a number of important metabolic enzymes in addition to ALAD (Chang et al., 2010, 2011; Scinicariello et al., 2010).

6   APPLICATIONS TO RISK ASSESSMENT PRACTICE

Based on the brief overview of how molecular biomarkers could be used for improving risk assessments, it is important to now examine some of the elements needed to translate this more basic information into risk assessment practice. In general terms, there are a number of data required. These include but are not limited to the following:

1. Solid and well-curated chemical exposure data which delineate specific chemicals, over a specified concentration range for populations of interest. Well-designed national biomonitoring studies such as the NHANES which address these fundamental issues are ideal for validating new molecular biomarkers in relation to specific chemical exposures and existing clinical endpoints.

2. Ancillary clinical data on the populations under study which provide basic information on age, gender, race/ethnicity, and genetic inheritance help to address the question of chemical exposures to whom? These data may expedite validation of MOA risk assessments by helping to identify subpopulations at special risk for adverse health outcomes and further refining interpretation of putative new molecular biomarkers in relation to measured standard clinical outcomes. In other words, sampling bias aside, these basic clinical data in identified populations at risk, may provide a best test case for both the validation and help in focusing evaluation of molecular biomarkers on productive outcomes.

3. Establishment of standard operating procedures (SOPs) with quality assurance/quality control (QA/QC) for all biomarker endpoint data used for risk assessment purposes. As with any data set used for societal decision making, it is important that all MOA risk assessment information be well-curated and reliable. This means that well-designed

QA/QC procedures for biomarker data should also be included at the start of any risk assessment project using these data.

4. Data analysis methods and approaches should be developed and specified prior to beginning a biomarker-based risk assessment project. These methods should be specified in the SOPs and include statistical methods to be employed for data analysis, power calculations, and expected summary statistical tables.

5. Since application of molecular biomarker endpoints for risk assessment is a relatively new and evolving field, a glossary of terms and brief explanation of the various methods employed would be of value for risk communication purposes to persons with limited technical backgrounds but who need to utilize the risk assessment information for decision making purposes.

References

Abe, M., Maruyama, N., Oikawa, O., Maruyama, T., Okada, K., Soma, M., 2015. Urinary ACE2 is associated with urinary L-FABP and albuminuria in patients with chronic kidney disease. Scand. J. Clin. Lab. Invest. 75 (5), 421–427.

Agrawal, S., Flora, G., Bhatnagar, P., Flora, S.J., 2014. Comparative oxidative stress, metallothionein induction and organ toxicity following chronic exposure to arsenic, lead and mercury in rats. Cell Mol. Biol. 60 (2), 13–21.

Akers, D.B., MacCarthy, M.F., Cunningham, J.A., Annis, J., Mihelcic, J.R., 2015. Lead (Pb) contamination of self-supply groundwater systems in coastal Madagascar and predictions of blood lead levels in exposed children. Environ. Sci. Technol. 49 (5), 2685–2693.

Bagrodia, A., Krabbe, L.M., Gayed, B.A., Kapur, P., Bernstein, I., Xie, X.J., et al., 2014. Evaluation of the prognostic significance of altered mammalian target of rapamycin pathway biomarkers in upper tract urothelial carcinoma. Urology 84 (5), 1134–1140.

Barfeld, S.J., East, P., Zuber, V., Mills, I.G., 2014. Meta-analysis of prostate cancer gene expression data identifies a novel discriminatory signature enriched for glycosylating enzymes. BMC Med. Genomics 7 (1), 513.

Becker, R.A., Ankley, G.T., Edwards, S.W., Kennedy, S.W., Linkov, I., Meek, B., et al., 2015. Increasing scientific confidence in adverse outcome pathways: application of tailored Bradford-Hill considerations for evaluating weight of evidence. Regul. Toxicol. Pharmacol. 72 (3), 514–537.

Bergdahl, I.A., Grubb, A., Schutz, A., Desnick, R.J., Wetmur, J.G., Sassa, S., Skerfving, S., 1997. Lead binding to delta-aminolevulinic acid dehydratase (ALAD) in human erythrocytes. Pharmacol. Toxicol. 81 (4), 153–158.

Burden, N., Sewell, F., Andersen, M.E., Boobis, A., Chipman, J.K., Cronin, M.T., et al., 2015. Adverse outcome pathways can drive non-animal approaches for safety assessment. J. Appl. Toxicol. 35, 971–975.

Candiano, G., Santucci, L., Petretto, A., Bruschi, M., Dimuccio, V., Urbani, A., et al., 2010. 2D-electrophoresis and the urine proteome map: where do we stand? J. Proteomics 73 (5), 829–844.

Carmichael, N., Fowler, B., 1981. Cadmium accumulation and toxicity in the kidney of the bay scallop *Argopecten irradians*. Mar. Biol. 65 (1), 35–43.

CDC., 2005. Third National Report on Human Exposure to Environmental Chemicals. US Department of Health and Human Services. Atlanta, GA. pp. 467.

Chang, M.H., Yesupriya, A., Ned, R.M., Mueller, P.W., Dowling, N.F., 2010. Genetic variants associated with fasting blood lipids in the U.S. population: Third National Health and Nutrition Examination Survey. BMC Med. Genet. 11, 62.

Chang, M.H., Ned, R.M., Hong, Y., Yesupriya, A., Yang, Q., Liu, T., et al., 2011. Racial/ethnic variation in the association of lipid-related genetic variants with blood lipids in the US adult population. Circ. Cardiovasc. Genet. 4 (5), 523–533.

Chen, M., Wang, J., 2002. Initiator caspases in apoptosis signaling pathways. Apoptosis 7 (4), 313–319.

Chen, J., Mercer, G., Roth, S.R., Abraham, L., Lutz, P., Ercal, N., Neal, R.E., 2011. Sub-chronic lead exposure alters kidney proteome profiles. Hum. Exp. Toxicol. 30 (10), 1616–1625.

Cogliano, V.J., Baan, R., Straif, K., Grosse, Y., Lauby-Secretan, B., El Ghissassi, F., et al., 2011. Preventable exposures associated with human cancers. J. Natl. Cancer Inst. 103 (24), 1827–1839.

Cong, N., Liu, L., Xie, Y., Shao, W., Song, J., 2014. Association between glutathione S-transferase T1, M1, and P1 genotypes and the risk of colorectal cancer. J. Korean Med. Sci. 29 (11), 1488–1492.

Conner, E.A., Fowler, B.A., 1994. Biochemical and immunological properties of hepatic δ-aminolevulinic acid dehydratase in channel catfish (*Ictalurus punctatus*). Aquat. Toxicol. 28 (1–2), 37–52.

Cote, I., Anastas, P.T., Birnbaum, L.S., Clark, R.M., Dix, D.J., Edwards, S.W., Preuss, P.W., 2012. Advancing the next generation of health risk assessment. Environ. Health Perspect. 120 (11), 1499–1502.

De Siervi, A., Vazquez, E.S., Rezaval, C., Rossetti, M.V., del Batlle, A.M., 2002. Delta-aminolevulinic acid cytotoxic effects on human hepatocarcinoma cell lines. BMC Cancer 2, 6.

Deng, Q., He, B., Pan, Y., Sun, H., Liu, X., Chen, J., et al., 2015. Polymorphisms of GSTA1 contribute to elevated cancer risk: evidence from 15 studies. J. BUON 20 (1), 287–295.

Economopoulos, K.P., Sergentanis, T.N., 2010. GSTM1, GSTT1, GSTP1, GSTA1 and colorectal cancer risk: a comprehensive meta-analysis. Eur. J. Cancer 46 (9), 1617–1631.

Etchevers, A., Le Tertre, A., Lucas, J.P., Bretin, P., Oulhote, Y., Le Bot, B., Glorennec, P., 2015. Environmental determinants of different blood lead levels in children: a quantile analysis from a nationwide survey. Environ. Int. 74, 152–159.

Fan, Y., Zhang, W., Shi, C.Y., Cai, D.F., 2013. Associations of GSTM1 and GSTT1 polymorphisms with pancreatic cancer risk: evidence from a meta-analysis. Tumour Biol. 34 (2), 705–712.

Feng, S., He, X., 2013. Mechanism-based inhibition of CYP450: an indicator of drug-induced hepatotoxicity. Curr. Drug Metab. 14 (9), 921–945.

Fletcher, M.E., Boshier, P.R., Wakabayashi, K., Keun, H.C., Smolenski, R.T., Kirkham, P.A., et al., 2015. Influence of glutathione-S-transferase (GST) inhibition on lung epithelial cell injury: role of oxidative stress and metabolism. Am. J. Physiol. Lung Cell Mol. Physiol. 308 (12), L1274–1285.

Fowler, B.A., 1980. Ultrastructural morphometric/biochemical assessment of cellular toxicity. In: Witschi, H.P. (Ed.), In: Proceedings of Symposium on The Scientific Basis of Toxicity Assessment. Elsevier Publishers, Amsterdam, pp. 211–218.

Fowler, B.A., 1983. Role of ultrastructural techniques in understanding mechanisms of metal-induced nephrotoxicity. Fed. Proc. 42 (13), 2957–2964.

Fowler, B.A., 2009. Monitoring of human populations for early markers of cadmium toxicity: a review. Toxicol. Appl. Pharmacol. 238 (3), 294–300.

Fowler, B.A., 2012. Biomarkers in toxicology and risk assessment. EXS 101, 459–470.

Fowler, B.A., 2013a. Cadmium and aging. In: Weiss, B. (Ed.), Aging and Vulnerabilities to Environmental Chemicals. Royal Society of Chemistry, Cambridge, UK, pp. 376–387.

Fowler, B.A. (Ed.), 2013b. Computational Toxicology: Methods and Applications for Risk Assessment. Elsevier Publishers, Amsterdam, p. 258.

Fowler, B.A., Mahaffey, K.R., 1978. Interactions among lead, cadmium, and arsenic in relation to porphyrin excretion patterns. Environ. Health Perspect. 25, 87–90.

Fowler, B.A., Nordberg, G.F., 1978. The renal toxicity of cadmium metallothionein: morphometric and X-ray microanalytical studies. Toxicol. Appl. Pharmacol. 46 (3), 609–623.

Fowler, B.A., Woods, J.S., 1977. The transplacental toxicity of methyl mercury to fetal rat liver mitochondria. Morphometric and biochemical studies. Lab. Invest. 36 (2), 122–130.

Fowler, B.A., Woods, J.S., 1979. The effects of prolonged oral arsenate exposure on liver mitochondria of mice: morphometric and biochemical studies. Toxicol. Appl. Pharmacol. 50 (2), 177–187.

Fowler, B.A., Brown, H.W., Lucier, G.W., Krigman, M.R., 1975. The effects of chronic oral methyl mercury exposure on the lysosome system of rat kidney. Morphometric and biochemical studies. Lab. Invest. 32 (3), 313–322.

Fowler, B.A., Woods, J.S., Schiller, C.M., 1979. Studies of hepatic mitochondrial structure and function: morphometric and biochemical evaluation of in vivo perturbation by arsenate. Lab. Invest. 41 (4), 313–320.

Fowler, B.A., Kimmel, C.A., Woods, J.S., McConnell, E.E., Grant, L.D., 1980. Chronic low-level lead toxicity in the rat. III. An integrated assessment of long-term toxicity with special reference to the kidney. Toxicol. Appl. Pharmacol. 56 (1), 59–77.

Fowler, B.A., Kardish, R.M., Woods, J.S., 1983. Alteration of hepatic microsomal structure and function by indium chloride. Ultrastructural, morphometric, and biochemical studies. Lab. Invest. 48 (4), 471–478.

Fowler, B.A., Hildebrand, C.E., Kojima, Y., Webb, M., 1987. Nomenclature of metallothionein. Experientia Suppl. 52, 19–22.

Fowler, B.A., Conner, E.A., Yamauchi, H., 2005. Metabolomic and proteomic biomarkers for III–V semiconductors: chemical-specific porphyrinurias and proteinurias. Toxicol. Appl. Pharmacol. 206 (2), 121–130.

Fowler, B.A., Conner, E.A., Yamauchi, H., 2008. Proteomic and metabolomic biomarkers for III–V semiconductors: and prospects for application to nano-materials. Toxicol. Appl. Pharmacol. 233 (1), 110–115.

Gallon, C., Ranville, M.A., Conaway, C.H., Landing, W.M., Buck, C.S., Morton, P.L., Flegal, A.R., 2011. Asian industrial lead inputs to the North Pacific evidenced by lead concentrations and isotopic compositions in surface waters and aerosols. Environ. Sci. Technol. 45 (23), 9874–9882.

Garcia-Vargas, G.G., Del Razo, L.M., Cebrian, M.E., Albores, A., Ostrosky-Wegman, P., Montero, R., et al., 1994. Altered urinary porphyrin excretion in a human population chronically exposed to arsenic in Mexico. Hum. Exp. Toxicol. 13 (12), 839–847.

Goering, P.L., Fowler, B.A., 1985. Mechanism of renal lead-binding protein reversal of delta-aminolevulinic acid dehydratase inhibition by lead. J. Pharmacol. Exp. Ther. 234 (2), 365–371.

Goering, P.L., Fowler, B.A., 1987a. Kidney zinc-thionein regulation of delta-aminolevulinic acid dehydratase inhibition by lead. Arch. Biochem. Biophys. 253 (1), 48–55.

Goering, P.L., Fowler, B.A., 1987b. Metal constitution of metallothionein influences inhibition of delta-aminolaevulinic acid dehydratase (porphobilinogen synthase) by lead. Biochem. J. 245 (2), 339–345.

Goering, P.L., Fowler, B.A., 1987c. Regulatory roles of high-affinity metal-binding proteins in mediating lead effects on delta-aminolevulinic acid dehydratase. Ann. NY Acad. Sci. 514, 235–247.

Goering, P.L., Mistry, P., Fowler, B.A., 1986. A low molecular weight lead-binding protein in brain attenuates lead inhibition of delta-aminolevulinic acid dehydratase: comparison with a renal lead-binding protein. J. Pharmacol. Exp. Ther. 237 (1), 220–225.

Grams, M.E., Sang, Y., Ballew, S.H., Gansevoort, R.T., Kimm, H., Kovesdy, C.P., et al., 2015. A meta-analysis of the association of estimated GFR, albuminuria, age, race, and sex with acute kidney injury. Am. J. Kidney Dis. 66, 591–601.

Grosse, Y., Loomis, D., Lauby-Secretan, B., El Ghissassi, F., Bouvard, V., Benbrahim-Tallaa, L., et al., 2013. Carcinogenicity of some drugs and herbal products. Lancet Oncol. 14 (9), 807–808.

Gulson, B., Anderson, P., Taylor, A., 2013. Surface dust wipes are the best predictors of blood leads in young children with elevated blood lead levels. Environ. Res. 126, 171–178.

Gulson, B., Mizon, K., Taylor, A., Korsch, M., Davis, J.M., Louie, H., et al., 2014. Pathways of Pb and Mn observed in a 5-year longitudinal investigation in young children and environmental measures from an urban setting. Environ. Pollut. 191, 38–49.

Han, Q.Y., Wang, H.X., Liu, X.H., Guo, C.X., Hua, Q., Yu, X.H., et al., 2015. Circulating E3 ligases are novel and sensitive biomarkers for diagnosis of acute myocardial infarction. Clin. Sci. 128 (11), 751–760.

Hegstad, A.C., Ytrehus, K., Myklebust, R., Jorgensen, L., 1994. Ultrastructural changes in the myo-cardial myocytic mitochondria: crucial step in the development of oxygen radical-induced damage in isolated rat hearts? Basic Res. Cardiol. 89 (2), 128–138.

Jaramillo-Rangel, G., Ortega-Martinez, M., Cerda-Flores, R.M., Barrera-Saldana, H.A., 2015. Short Communication Polymorphisms in GSTM1, GSTT1, GSTP1, and GSTM3 genes and breast cancer risk in northeastern Mexico. Genet. Mol. Res. 14 (2), 6465–6471.

Joshi, S., Viljoen, A., 2015. Renal biomarkers for the prediction of cardiovascular disease. Curr. Opin. Cardiol. 30 (4), 454–460.

Kassogue, Y., Dehbi, H., Quachouh, M., Quessar, A., Benchekroun, S., Nadifi, S., 2015. Associa-tion of glutathione S-transferase (GSTM1 and GSTT1) genes with chronic myeloid leukemia. Springerplus 4, 210.

Kataria, A., Trasande, L., Trachtman, H., 2015. The effects of environmental chemicals on renal function. Nat. Rev. Nephrol. 11, 610–625.

Kim, H.J., Ha, S., Lee, H.Y., Lee, K.J., 2015. ROSics: chemistry and proteomics of cysteine modifica-tions in redox biology. Mass Spectrom. Rev. 34 (2), 184–208.

Klawitter, J., Klawitter, J., Kushner, E., Jonscher, K., Bendrick-Peart, J., Leibfritz, D., et al., 2010. Association of immunosuppressant-induced protein changes in the rat kidney with changes in urine metabolite patterns: a proteo-metabonomic study. J. Proteome Res. 9 (2), 865–875.

Kohane, I.S., 2015. Health Care Policy. Ten things we have to do to achieve precision medicine. Science 349 (6243), 37–38.

Kruger, M., Pabst, A.M., Mahmoodi, B., Becker, B., Kammerer, P.W., Koch, F.P., 2015. The impact of GSTM1/GSTT1 polymorphism for the risk of oral cancer. Clin. Oral Investig. 19, 1791–1797.

Laidlaw, M.A., Zahran, S., Pingitore, N., Clague, J., Devlin, G., Taylor, M.P., 2014. Identification of lead sources in residential environments: Sydney Australia. Environ. Pollut. 184, 238–246.

Lee, D.H., Lee, I.K., Song, K., Steffes, M., Toscano, W., Baker, B.A., Jacobs, Jr., D.R., 2006. A strong dose-response relation between serum concentrations of persistent organic pollutants and dia-betes: results from the National Health and Examination Survey 1999–2002. Diabetes Care 29 (7), 1638–1644.

Lee, D.H., Lee, I.K., Jin, S.H., Steffes, M., Jacobs, Jr., D.R., 2007. Association between serum con-centrations of persistent organic pollutants and insulin resistance among nondiabetic adults: results from the National Health and Nutrition Examination Survey 1999–2002. Diabetes Care 30 (3), 622–628.

Mahaffey, K.R., Capar, S.G., Gladen, B.C., Fowler, B.A., 1981. Concurrent exposure to lead, cadmium, and arsenic. Effects on toxicity and tissue metal concentrations in the rat. J. Lab. Clin. Med. 98 (4), 463–481.

Mahaffey, K.R., Clickner, R.P., Jeffries, R.A., 2009. Adult women's blood mercury concentrations vary regionally in the United States: association with patterns of fish consumption (NHANES 1999–2004). Environ. Health Perspect. 117 (1), 47–53.

Meek, M.E., Boobis, A., Cote, I., Dellarco, V., Fotakis, G., Munn, S., et al., 2014. New developments in the evolution and application of the WHO/IPCS framework on mode of action/species concordance analysis. J. Appl. Toxicol. 34 (1), 1–18.

NAS/NRC, 1993. Measuring Lead Exposure in Infants, Children and Other Sensitive Populations. National Academies Press, Washington, DC, pp. 337.

National Toxicology Program, 2004. NTP 11th Report on Carcinogens. Rep. Carcinog. 11, 1-A32.

National Toxicology Program, 2011. NTP 12th Report on Carcinogens. Rep. Carcinog. 12, iii-499.

Ngueta, G., Prevost, M., Deshommes, E., Abdous, B., Gauvin, D., Levallois, P., 2014. Exposure of young children to household water lead in the Montreal area (Canada): the potential influence of winter-to-summer changes in water lead levels on children's blood lead concentration. Environ. Int. 73, 57–65.

Ngueta, G., Abdous, B., Tardif, R., St-Laurent, J., Levallois, P., 2015. Use of a cumulative exposure index to estimate the impact of tap-water lead concentration on blood lead levels in 1- to 5-year-old children (Montreal, Canada). Environ. Health Perspect. 124, 388–395.

Orrenius, S., Zhivotovsky, B., Nicotera, P., 2003. Regulation of cell death: the calcium–apoptosis link. Nat. Rev. Mol. Cell Biol. 4 (7), 552–565.

Ott, M., Gogvadze, V., Orrenius, S., Zhivotovsky, B., 2007. Mitochondria, oxidative stress and cell death. Apoptosis 12 (5), 913–922.

Pallet, N., Chauvet, S., Chasse, J.F., Vincent, M., Avillach, P., Levi, C., et al., 2014. Urinary retinol binding protein is a marker of the extent of interstitial kidney fibrosis. PLoS One 9 (1), e84708.

Panotopoulos, J., Posch, F., Alici, B., Funovics, P., Stihsen, C., Amann, G., et al., 2015. Hemoglobin, alkalic phosphatase, and C-reactive protein predict the outcome in patients with liposarcoma. J. Orthop. Res. 33 (5), 765–770.

Phillips, M.B., Leonard, J.A., Grulke, C.M., Chang, D.T., Edwards, S.W., Brooks, R., et al., 2015. A workflow to investigate exposure and pharmacokinetic influences on high-throughput chemical screening based on adverse outcome pathways. Environ. Health Perspect. 124, 53–60.

Pirkle, J.L., Brody, D.J., Gunter, E.W., Kramer, R.A., Paschal, D.C., Flegal, K.M., Matte, T.D., 1994. The decline in blood lead levels in the United States. The National Health and Nutrition Examination Surveys (NHANES). JAMA 272 (4), 284–291.

Pirkle, J.L., Kaufmann, R.B., Brody, D.J., Hickman, T., Gunter, E.W., Paschal, D.C., 1998. Exposure of the U.S. population to lead, 1991–1994. Environ. Health Perspect. 106 (11), 745–750.

Quintanilla-Vega, B., Smith, D.R., Kahng, M.W., Hernandez, J.M., Albores, A., Fowler, B.A., 1995. Lead-binding proteins in brain tissue of environmentally lead-exposed humans. Chem. Biol. Interact. 98 (3), 193–209.

Ruiz, P., Mumtaz, M., Osterloh, J., Fisher, J., Fowler, B.A., 2010. Interpreting NHANES biomonitoring data, cadmium. Toxicol. Lett. 198 (1), 44–48.

Ryter, S.W., Choi, A.M., 2013. Regulation of autophagy in oxygen-dependent cellular stress. Curr. Pharm. Des. 19 (15), 2747–2756.

Ryter, S.W., Tyrrell, R.M., 2000. The heme synthesis and degradation pathways: role in oxidant sensitivity. Heme oxygenase has both pro- and antioxidant properties. Free Radic. Biol. Med. 28 (2), 289–309.

Saikumar, P., Dong, Z., Mikhailov, V., Denton, M., Weinberg, J.M., Venkatachalam, M.A., 1999. Apoptosis: definition, mechanisms, and relevance to disease. Am. J. Med. 107 (5), 489–506.

Scinicariello, F., Yesupriya, A., Chang, M.H., Fowler, B.A., 2010. Modification by ALAD of the association between blood lead and blood pressure in the U.S. population: results from the Third National Health and Nutrition Examination Survey. Environ. Health Perspect. 118 (2), 259–264.

Shin, J.R., Kim, S.M., Yoo, J.S., Park, J.Y., Kim, S.K., Cho, J.H., et al., 2014. Urinary excretion of beta2-microglobulin as a prognostic marker in immunoglobulin A nephropathy. Korean J. Intern. Med. 29 (3), 334–340.

Smith, D.R., Kahng, M.W., Quintanilla-Vega, B., Fowler, B.A., 1998. High-affinity renal lead-binding proteins in environmentally-exposed humans. Chem. Biol. Interact. 115 (1), 39–52.

Song, K., Yi, J., Shen, X., Cai, Y., 2012. Genetic polymorphisms of glutathione S-transferase genes GSTM1, GSTT1 and risk of hepatocellular carcinoma. J. Korean Med. Sci. 7 (11), e48924.

Stacchiotti, A., Lavazza, A., Rezzani, R., Bianchi, R., 2002. Cyclosporine A-induced kidney alterations are limited by melatonin in rats: an electron microscope study. Ultrastruct. Pathol. 26 (2), 81–87.

Straif, K., Benbrahim-Tallaa, L., Baan, R., Grosse, Y., Secretan, B., El Ghissassi, F., et al., 2009. A review of human carcinogens—Part C: metals, arsenic, dusts, and fibres. Lancet Oncol. 10 (5), 453–454.

Tan, M.C., Basturk, O., Brannon, A.R., Bhanot, U., Scott, S.N., Bouvier, N., et al., 2015. GNAS and KRAS mutations define separate progression pathways in intraductal papillary mucinous neoplasm-associated carcinoma. J. Am. Coll. Surg. 220 (5), 845.e1–854.e1.

Taylor, K.W., Novak, R.F., Anderson, H.A., Birnbaum, L.S., Blystone, C., Devito, M., et al., 2013. Evaluation of the association between persistent organic pollutants (POPs) and diabetes in epidemiological studies: a national toxicology program workshop review. Environ. Health Perspect. 121 (7), 774–783.

Tellez-Plaza, M., Navas-Acien, A., Caldwell, K.L., Menke, A., Muntner, P., Guallar, E., 2012a. Reduction in cadmium exposure in the United States population, 1988–2008: the contribution of declining smoking rates. Environ. Health Perspect. 120 (2), 204–209.

Tellez-Plaza, M., Navas-Acien, A., Menke, A., Crainiceanu, C.M., Pastor-Barriuso, R., Guallar, E., 2012b. Cadmium exposure and all-cause and cardiovascular mortality in the U.S. general population. Environ. Health Perspect. 120 (7), 1017–1022.

Tian, L., Zheng, G., Sommar, J.N., Liang, Y., Lundh, T., Broberg, K., et al., 2013. Lead concentration in plasma as a biomarker of exposure and risk, and modification of toxicity by delta-aminolevulinic acid dehydratase gene polymorphism. Toxicol. Lett. 221 (2), 102–109.

Tikoo, K., Patel, G., Kumar, S., Karpe, P.A., Sanghavi, M., Malek, V., Srinivasan, K., 2015. Tissue specific up regulation of ACE2 in rabbit model of atherosclerosis by atorvastatin: role of epigenetic histone modifications. Biochem. Pharmacol. 93 (3), 343–351.

Varghese, S.A., Powell, T.B., Janech, M.G., Budisavljevic, M.N., Stanislaus, R.C., Almeida, J.S., Arthur, J.M., 2010. Identification of diagnostic urinary biomarkers for acute kidney injury. J. Investig. Med. 58 (4), 612–620.

Vinken, M., 2015. Adverse outcome pathways and drug-induced liver injury testing. Chem. Res. Toxicol. 28, 1391–1397.

Whittaker, M.H., Wang, G., Chen, X.Q., Lipsky, M., Smith, D., Gwiazda, R., Fowler, B.A., 2011. Exposure to Pb, Cd, and As mixtures potentiates the production of oxidative stress precursors: 30-day, 90-day, and 180-day drinking water studies in rats. Toxicol. Appl. Pharmacol. 254 (2), 154–166.

Wilson, J., Dixon, S.L., Jacobs, D.E., Akoto, J., Korfmacher, K.S., Breysse, J., 2015. An investigation into porch dust lead levels. Environ. Res. 137, 129–135.

Woods, J.S., Fowler, B.A., 1978. Altered regulation of mammalian hepatic heme biosynthesis and urinary porphyrin excretion during prolonged exposure to sodium arsenate. Toxicol. Appl. Pharmacol. 43 (2), 361–371.

Woods, J.S., Fowler, B.A., 1986. Alteration of hepatocellular structure and function by thallium chloride: ultrastructural, morphometric, and biochemical studies. Toxicol. Appl. Pharmacol. 83 (2), 218–229.

Wu, R.T., Cao, L., Chen, B.P., Cheng, W.H., 2014. Selenoprotein H suppresses cellular senescence through genome maintenance and redox regulation. J. Biol. Chem. 289 (49), 34378–34388.

Wu, G., Huang, S., Nastiuk, K.L., Li, J., Gu, J., Wu, M., et al., 2015. Variant allele of HSD3B1 increases progression to castration-resistant prostate cancer. Prostate 75 (7), 777–782.

Wunnapuk, K., Liu, X., Peake, P., Gobe, G., Endre, Z., Grice, J.E., et al., 2013. Renal biomarkers predict nephrotoxicity after paraquat. Toxicol. Lett. 222 (3), 280–288.

Yauk, C.L., Lambert, I.B., Meek, M.E., Douglas, G.R., Marchetti, F., 2015. Development of the adverse outcome pathway "alkylation of DNA in male premeiotic germ cells leading to heritable mutations" using the OECD's users' handbook supplement. Environ. Mol. Mutagen. 56, 724–750.

Yeh, H.C., Lin, Y.S., Kuo, C.C., Weidemann, D., Weaver, V., Fadrowski, J., et al., 2015. Urine osmolality in the US population: implications for environmental biomonitoring. Environ. Res. 136, 482–490.

Yoon, H.J., Lee, Y.H., Kim, S.R., Rim, T.H., Lee, E.Y., Kang, E.S., et al., 2015. Glycated albumin and the risk of micro- and macrovascular complications in subjects with type 1 diabetes. Cardiovasc. Diabetol. 14, 53.

Zheng, L., Kuo, C.C., Fadrowski, J., Agnew, J., Weaver, V.M., Navas-Acien, A., 2014. Arsenic and chronic kidney disease: a systematic review. Curr. Environ. Health Rep. 1 (3), 192–207.

CHAPTER 6

Technical Translational Analysis of Molecular Biomarker Data

1 INTRODUCTION

1.1 Data Analysis

As with any type of data, appropriate analysis of measured changes in molecular biomarkers endpoints is essential for making accurate interpretations of the findings. These analyses may employ statistical or biological models or a combination of both. On the one hand, it is possible that results found to be statistically significant may be of such a small magnitude as to be of limited biological consequence. On the other hand, statistical analyses are extremely valuable for objective evaluations of apparent changes in a biomarker endpoint and to confirm associations between chemical exposures and biomarker responses (Hack et al., 2010). Ultimately sound judgment on the part of the risk assessor is needed on how best to effectively apply these tools to reach wise decisions.

1.2 Types of Statistical Analyses

There are many papers and books written on statistical tests and many available computer programs designed to aid investigators in evaluating data sets. The important component in the evaluation process is to apply the most appropriate statistical methodology for evaluating the biological problem of interest, given the data that are available. (Burke et al., 2015; Ding et al., 2014; Mendell et al., 2015; Rikke et al., 2015; Samavat et al., 2015). Examples of some commercially available statistical analysis packages are presented in Table 6.1. This is not an inclusive listing and there are many more such programs with specific applications.

1.3 Evaluation of Statistical Analyses

As noted earlier, while it is important to have sound statistical analyses using appropriate tests, it also important to stand back and look at the findings using a Bradford Hill criteria (Hill, 1965) approach, that is, take into account the

CONTENTS

Molecular Biological Markers for Toxicology and Risk Assessment. http://dx.doi.org/10.1016/B978-0-12-809589-8.00006-8

Table 6.1 Commercially Available Statistical Analysis Packages of Value for Analysis of Validating Molecular Biomarker Data Sets

Vendor	Type of Analysis	Value of Output for Validation Purposes
SAS Institute	Multiple regression	Statistical correlation[a]
Analytica	Modeling for risk analysis	Decision analysis
Genstat	Bioscience modeling	Statistical analysis
MedCalc	Biomedical applications	Statistical analysis
SPSS	General statistical analysis	Statistical analysis
MatLab	Mathematics	Statistical analysis
Gene Data Analyst	Analysis and visualization of large bioscience data sets	Analysis of large bioscience data sets

[a]Between molecular biomarker and other health outcome(s).

existing state of knowledge. Are the findings biologically plausible and significant and do they make sense in light of other relevant data in the study and data from other studies? The overall point being not to simply accept the results of the statistical analyses at face value but to consider their potential biological significance and their coherence. In this way alternative interpretations and possible health outcomes should emerge and be given serious consideration.

2 MODELING AND INTERPRETATION OF DATA

Once one is satisfied with the quality of the data and the results of statistical and biological evaluation, it is then important to consider how best to translate the analyzed data for risk assessment purposes. This is commonly done in other fields through first developing predictive models and then testing the models. This provides assurance that the findings are being correctly interpreted and that conclusions can be drawn with greater confidence.

2.1 Model Development

2.1.1 Evaluating the Model

A predictive model may be developed by using available data sets from the literature or unpublished available laboratory data. In the case of the latter or any published data that have not been peer-reviewed, it is important to perform appropriate QA/QC evaluations to the greatest extent possible. Once an appropriate model has been developed, it is important to *refine* it by repetitive testing through incorporation of related similar chemical exposure data and retesting the model. Ideally, this would be done multiple times. These steps will refine the model and make it more robust since the expanded data set upon which the model is based is increased and hence more complete. This approach is used extensively in modern model development in fields such as QSAR (Demchuk et al., 2011) and PBPK (Ruiz et al., 2010).

2.1.2 Testing the Model

Once the model has been developed, it is important to determine its predictive accuracy by comparing the projected predictions with actual data of interest. One example of this is the study by Ruiz et al. (2010), which compared Berkeley–Madonna predictions for urinary cadmium concentrations for the general US population with actual analytical data for urinary cadmium values from persons participating in the NHANESIII (CDC, 2005).

The results were in close agreement across a wide cadmium exposure range indicating that the model could be used for predicting which groups within the general population would be expected to be at greatest risk for cadmium toxicity.

3 INTEGRATION OF DIVERSE DATA SETS

It is clear that in order to be useful for risk assessment purposes, any modeling approach must be able to link dose/exposure data to health outcome effect data in a credible manner. This means that in most cases—the exception being large composite studies, such as National Health and Nutrition Examination Survey (NHANES)—modeling efforts must be able to integrate different types of data (eg, analytical chemical exposure data, molecular biomarker data, genetic, nutritional, and demographic data) from multiple sources and generated for different purposes in order to derive the best possible risk assessment. This is clearly a complex but not impossible task if one looks for ways to standardize the data so that can be aligned with each other. Standardization of data is possible as a transparent exercise if one provides the adjustment factors employed and justifies them on a scientific basis. For example, this approach has been used for relating the relative toxicity of a number of chlorinated or brominated organic compounds commonly found as mixtures through the use of a toxicity equivalency factor (TEF) which relates the chemical toxicity of the chemicals in a mixture to that of dioxin (Bhavsar et al., 2008; Chao et al., 2006; Hong et al., 2009; Rigaud et al., 2014; Sutter et al., 2006; van den Berg et al., 2013). It has also been appreciated that other types of chemicals, such as arsenic may alter the expected metabolism of chlorinated organics (Chao et al., 2006). Species differences in sensitivity to chlorinated organics (Sutter et al., 2006) may also alter the precision of a calculated TEF. These types of interactive factors, when present, may alter expected TEF-based risk assessments. The point here is that the TEF approach for risk assessment purposes may also be subject to interactions from other common chemical exposures; so such possible interactions should also be considered in an overall risk assessment.

4   VALIDATION OF BIOLOGICAL MARKER DATA WITH OTHER OUTCOME DATA

As noted earlier, it is essential that putative molecular biomarker data are validated in relation to other types of analytical, clinical, genetic, and health outcome data in order to be of maximal value for risk assessment purposes. In published studies a variety of statistical approaches for validating biomarkers in relation to a number of health conditions can be found (Obuchowski et al., 2015; Wang et al., 2013; Xi et al., 2014). Caveats noted in the incorporation of prefiltering statistical algorithms to small data sets prior to analysis may obfuscate results due to overfitting (Hernandez et al., 2014). In addition, Bayesian analysis methods represent another approach to computational model validation (Bois et al., 1996; Cai et al., 2014; Hack et al., 2010; Jiang et al., 2014a,b, 2015; Jiang and Neapolitan, 2015; Mahadevan and Rebba, 2005; Neapolitan et al., 2014). Merging disparate types of data for analysis is a complex task that requires innovative thinking and the use of sophisticated computer modeling. However, with the advances made in relevant scientific areas, it is an approach that will yield improved risk assessments. A major goal is for molecular biomarkers to become accepted as valuable tools for risk assessment practice. This is of critical importance since any societal decision based upon the molecular biomarker approaches outlined earlier will need to stand up to intense scientific and legalistic scrutiny.

4.1   Other Potentially Useful Types of Correlative Health Outcome Data for Biological Marker Validation

Another potentially useful approach is to mine publically available databases to look for linkages between biomarkers measured on archived samples of blood, urine, or other matrices and clinical endpoints such as standard serum enzyme activities or clinical diseases like diabetes. In this way, the putative biomarker could be biologically validated in a clinical setting. The following are possible examples of where this proposed approach could be tested.

4.1.1   NHANES Database

The NHANES is a very large database which started in the 1970s to identify relationships between nutritional status and overall health in a representative sample of the general US population from ages 1 to 75 years. It is a very comprehensive and well-curated database which has been expanded over the years to include measurements of a number of toxic chemical agents. It is now possible to evaluate trends in both chemical exposures and possible linkages to health outcomes.(Garcia-Esquinas et al., 2015; Lee et al., 2006, 2007, 2010, 2014; Menke et al., 2014; Wiener et al., 2015).

4.1.2 European Union Health Databases

Similar types of health data are available for countries in the European Union (EU) but on a more country-by-country basis (Berghofer et al., 2008; Gerth et al., 2002; Varo et al., 2003). Nonetheless, there is an increasing number of curated databases in EU countries which contain potentially valuable health information which could be mined to provide correlative data for validation of molecular biomarker endpoints. It would be very interesting to conduct meta-analyses for evolving molecular biomarkers that utilize available public health databases from the United States, EU, and Asian countries, such as Korea, to strengthen validation conclusions for molecular biomarkers from a global perspective.

4.1.3 Korean NHANES

The Korean NHANES program (Khang and Yun, 2010; Seo et al., 2015) provides national Korean health data similar to the US NHANES. Increasingly countries are developing correlative national health data sets which could also be mined to provide validation information for molecular biomarker studies. Such data represent potentially very valuable information as the mean age of populations in many countries continues to increase, and geriatric populations may be regarded as a population at special risk for chemical toxicity. In addition, molecular biomarkers may also be used to help monitor and manage the roles of chemicals in the aging process. One example in this regard is that of cadmium (Fowler, 2013).

4.1.4 Clinical Chemistry Analyzer Databases

With the advent of automated clinical chemistry analyzers which simultaneously measure multiple analytes, there is now a large commercial database which contains the information on the normal range of values for persons undergoing routine clinical examinations. This provides the opportunity for comparing findings from putative molecular biomarker tests with more standard and accepted clinical endpoints. Since these automated clinical chemistry analyzers are now managed by computer systems it should be relatively easy to eventually include omic tests in the analyses. The direct comparative evaluations of putative biomarkers for validation purposes would greatly expedite the acceptance of new biomarker tests.

4.2 Utilization of Molecular Biomarkers as Translational Bridges Between Epidemiological and Standard Clinical Chemistry Data Sets

Molecular biomarkers are useful as "translational biomarkers" since they can provide mechanistic bridges between associative epidemiological data and standard clinical laboratory data. Mechanistic linkages between these major

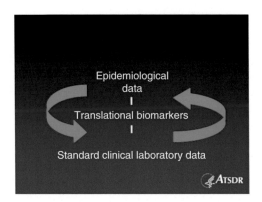

FIGURE 6.1 Conceptual linkages by which molecular biomarkers could be used to provide a mechanistic bridge between epidemiological data and results of standard clinical laboratory measurements.

types of data are essential for making a rigorous case for causality. Fig. 6.1 highlights the conceptual linkages. These types of connections are also clearly essential for utilization of epidemiological and clinical chemistry data in the development of actual MOA strategies for risk assessment practice.

References

Berghofer, A., Pischon, T., Reinhold, T., Apovian, C.M., Sharma, A.M., Willich, S.N., 2008. Obesity prevalence from a European perspective: a systematic review. BMC Public Health 8, 200.

Bhavsar, S.P., Reiner, E.J., Hayton, A., Fletcher, R., MacPherson, K., 2008. Converting toxic equivalents (TEQ) of dioxins and dioxin-like compounds in fish from one toxic equivalency factor (TEF) scheme to another. Environ. Int. 34 (7), 915–921.

Bois, F.Y., Gelman, A., Jiang, J., Maszle, D.R., Zeise, L., Alexeef, G., 1996. Population toxicokinetics of tetrachloroethylene. Arch. Toxicol. 70 (6), 347–355.

Burke, O., Benton, S., Szafranski, P., von Dadelszen, P., Buhimschi, C., Cetin, I., et al., 2015. [94-OR]: extending the scope of individual patient data meta-analyses: Merging algorithms for biomarker measurements from heterogeneous laboratory platforms. The CoLAB Preeclampsia angiogenic factor study. Pregnancy Hypertens., 5(1), 50-51.

Cai, C., Chen, L., Jiang, X., Lu, X., 2014. Modeling signal transduction from protein phosphorylation to gene expression. Cancer Inform. 13 (Suppl. 1), 59–67.

CDC, 2005. Third national report on human exposure to environmental chemicals. National Center for Environmental Health, NCEH Pub. 05-0570.

Chao, H.R., Tsou, T.C., Li, L.A., Tsai, F.Y., Wang, Y.F., Tsai, C.H., et al., 2006. Arsenic inhibits induction of cytochrome P450 1A1 by 2, 3,7,8-tetrachlorodibenzo-*p*-dioxin in human hepatoma cells. J. Hazard Mater. 137 (2), 716–722.

Demchuk, E., Ruiz, P., Chou, S., Fowler, B.A., 2011. SAR/QSAR methods in public health practice. Toxicol. Appl. Pharmacol. 254 (2), 192–197.

Ding, Y.C., Yu, W., Ma, C., Wang, Q., Huang, C.S., Huang, T., 2014. Expression of long non-coding RNA LOC285194 and its prognostic significance in human pancreatic ductal adenocarcinoma. Int. J. Clin. Exp. Pathol. 7 (11), 8065–8070.

Fowler, B.A., 2013. Cadmium and aging. In: Weiss, B. (Ed.), Aging and Vulnerabilities to Environmental Chemicals. Royal Society of Chemistry, Cambridge, UK, pp. 376–387.

Garcia-Esquinas, E., Navas-Acien, A., Perez-Gomez, B., Artalejo, F.R., 2015. Association of lead and cadmium exposure with frailty in US older adults. Environ. Res. 137, 424–431.

Gerth, W.C., Remuzzi, G., Viberti, G., Hannedouche, T., Martinez-Castelao, A., Shahinfar, S., et al., 2002. Losartan reduces the burden and cost of ESRD: public health implications from the RENAAL study for the European Union. Kidney Int. 62 (S82), S68–S72.

Hack, C.E., Haber, L.T., Maier, A., Shulte, P., Fowler, B., Lotz, W.G., Savage, Jr., R.E., 2010. A Bayesian network model for biomarker-based dose response. Risk Anal. 30 (7), 1037–1051.

Hernandez, B., Parnell, A., Pennington, S.R., 2014. Why have so few proteomic biomarkers "survived" validation? (Sample size and independent validation considerations). Proteomics 14 (13–14), 1587–1592.

Hill, A.B., 1965. The environment and disease: association or causation? Proc. Royal Soc. Med. 58 (5), 295–300.

Hong, B., Garabrant, D., Hedgeman, E., Demond, A., Gillespie, B., Chen, Q., et al., 2009. Impact of WHO 2005 revised toxic equivalency factors for dioxins on the TEQs in serum, household dust and soil. Chemosphere 76 (6), 727–733.

Jiang, X., Neapolitan, R.E., 2015. LEAP: biomarker inference through learning and evaluating association patterns. Genet. Epidemiol. 39 (3), 173–184.

Jiang, Y., Boyle, D.K., Bott, M.J., Wick, J.A., Yu, Q., Gajewski, B.J., 2014a. Expediting clinical and translational research via Bayesian instrument development. Appl. Psychol. Meas. 38 (4), 296–310.

Jiang, Z., Song, Y., Shou, Q., Xia, J., Wang, W., 2014b. A Bayesian prediction model between a biomarker and the clinical endpoint for dichotomous variables. Trials 15, 500.

Jiang, J., Zhang, Q., Ma, L., Li, J., Wang, Z., Liu, J.F., 2015. Joint prediction of multiple quantitative traits using a Bayesian multivariate antedependence model. Heredity 115 (1), 29–36.

Khang, Y.-H., Yun, S.-C., 2010. Trends in general and abdominal obesity among Korean adults: findings from 1998, 2001, 2005, and 2007 Korea National Health and Nutrition Examination Surveys. J. Korean Med. Sci. 25 (11), 1582–1588.

Lee, D.H., Lee, I.K., Song, K., Steffes, M., Toscano, W., Baker, B.A., Jacobs, Jr., D.R., 2006. A strong dose-response relation between serum concentrations of persistent organic pollutants and diabetes: results from the National Health and Examination Survey 1999–2002. Diabetes Care 29 (7), 1638–1644.

Lee, D.H., Lee, I.K., Jin, S.H., Steffes, M., Jacobs, Jr., D.R., 2007. Association between serum concentrations of persistent organic pollutants and insulin resistance among nondiabetic adults: results from the National Health and Nutrition Examination Survey 1999–2002. Diabetes Care 30 (3), 622–628.

Lee, D.H., Steffes, M.W., Sjodin, A., Jones, R.S., Needham, L.L., Jacobs, Jr., D.R., 2010. Low dose of some persistent organic pollutants predicts type 2 diabetes: a nested case-control study. Environ. Health Perspect. 118 (9), 1235–1242.

Lee, D.H., Porta, M., Jacobs, Jr., D.R., Vandenberg, L.N., 2014. Chlorinated persistent organic pollutants, obesity, and type 2 diabetes. Endocr. Rev. 35 (4), 557–601.

Mahadevan, S., Rebba, R., 2005. Validation of reliability computational models using Bayes networks. Reliab. Eng. Syst. Saf. 87 (2), 223–232.

Mendell, J., Freeman, D.J., Feng, W., Hettmann, T., Schneider, M., Blum, S., et al., 2015. Clinical translation and validation of a predictive biomarker for patritumab, an anti-human epidermal growth factor receptor 3 (HER3) monoclonal antibody, in patients with advanced non-small cell lung cancer. EBioMedicine 2 (3), 264–271.

Menke, A., Casagrande, S.S., Cowie, C.C., 2014. The relationship of adiposity and mortality among people with diabetes in the US general population: a prospective cohort study. BMJ Open 4 (11), e005671.

Neapolitan, R., Xue, D., Jiang, X., 2014. Modeling the altered expression levels of genes on signaling pathways in tumors as causal Bayesian networks. Cancer Inform. 13, 77–84.

Obuchowski, N.A., Reeves, A.P., Huang, E.P., Wang, X.F., Buckler, A.J., Kim, H.J., et al., 2015. Quantitative imaging biomarkers: a review of statistical methods for computer algorithm comparisons. Stat. Methods Med. Res. 24 (1), 68–106.

Rigaud, C., Couillard, C.M., Pellerin, J., Legare, B., Hodson, P.V., 2014. Applicability of the TCDD-TEQ approach to predict sublethal embryotoxicity in Fundulus heteroclitus. Aquat. Toxicol. 149, 133–144.

Rikke, B.A., Wynes, M.W., Rozeboom, L.M., Baron, A.E., Hirsch, F.R., 2015. Independent validation test of the vote-counting strategy used to rank biomarkers from published studies. Biomark. Med. 9, 1–11.

Ruiz, P., Fowler, B.A., Osterloh, J.D., Fisher, J., Mumtaz, M., 2010. Physiologically based pharmacokinetic (PBPK) tool kit for environmental pollutants—metals. SAR QSAR Environ. Res. 21 (7–8), 603–618.

Samavat, H., Dostal, A.M., Wang, R., Bedell, S., Emory, T.H., Ursin, G., et al., 2015. The Minnesota Green Tea Trial (MGTT), a randomized controlled trial of the efficacy of green tea extract on biomarkers of breast cancer risk: study rationale, design, methods, and participant characteristics. Cancer Causes Control. 26, 1405–1419.

Seo, J.W., Kim, B.G., Kim, Y.M., Kim, R.B., Chung, J.Y., Lee, K.M., Hong, Y.S., 2015. Trend of blood lead, mercury, and cadmium levels in Korean population: data analysis of the Korea National Health and Nutrition Examination Survey. Environ. Monit. Assess. 187 (3), 146.

Sutter, C.H., Rahman, M., Sutter, T.R., 2006. Uncertainties related to the assignment of a toxic equivalency factor for 1,2,3,4,6,7,8,9-octachlorodibenzo-p-dioxin. Regul. Toxicol. Pharmacol. 44 (3), 219–225.

van den Berg, M., Denison, M.S., Birnbaum, L.S., Devito, M.J., Fiedler, H., Falandysz, J., et al., 2013. Polybrominated dibenzo-p-dioxins, dibenzofurans, and biphenyls: inclusion in the toxicity equivalency factor concept for dioxin-like compounds. Toxicol. Sci. 133 (2), 197–208.

Varo, J.J., Martínez-González, M.A., de Irala-Estévez, J., Kearney, J., Gibney, M., Martínez, J.A., 2003. Distribution and determinants of sedentary lifestyles in the European Union. Int. J. Epidemiol. 32 (1), 138–146.

Wang, M., Mehta, A., Block, T.M., Marrero, J., Di Bisceglie, A.M., Devarajan, K., 2013. A comparison of statistical methods for the detection of hepatocellular carcinoma based on serum biomarkers and clinical variables. BMC Med. Genomics 6 (Suppl. 3), S9.

Wiener, R.C., Long, D.L., Jurevic, R.J., 2015. Blood levels of the heavy metal, lead, and caries in children aged 24-72 months: NHANES III. Caries Res. 49 (1), 26–33.

Xi, B., Gu, H., Baniasadi, H., Raftery, D., 2014. Statistical analysis and modeling of mass spectrometry-based metabolomics data. Methods Mol. Biol. 1198, 333–353.

CHAPTER 7

Quality Assurance/Quality Control (QA/QC) for Biomarker Data Sets

1 INTRODUCTION

In order for molecular biomarkers to again acceptance as trusted tools for risk assessment, they must meet the same scientific standards of rigor as other toxicological endpoints. This means that they must be technically validated by transparent quality assurance/quality control (QA/QC) procedures (Laurie et al., 2010; Wagner, 1995; Whitney et al., 1998). The value of including validation procedures for molecular biomarkers and the application of appropriate statistical analyses in order to warrant their incorporation in risk assessment has been recently reviewed by DeBord et al. (2015). It should be noted that such validation efforts may be a bit challenging since molecular biomarkers (eg, omics-based) are relatively new and rapidly evolving. An early step is a thoughtful selection of a prospective study design prior to incorporating a given biomarker as an endpoint in risk assessment. The general principles of QA/QC must also be included into study protocols at early stages of development, and achieving this will take resources (Fig. 7.1). Nonetheless, the playoff in terms of speed, reduced costs, and increased precision for mode of action (MOA) risk assessments justify the additional effort. Further, the credibility and transparency of the biomarker data results from a solid QA/QC program increases the probability that good risk assessment decisions are made and accepted.

The following are some short general definitions of QA/QC originally derived from the Merriam–Webster Dictionary definitions but now oriented toward molecular biomarkers. The overall purpose of these validation activities, in the context of molecular biomarkers, is to maintain the best possible quality for any biomarker so that the data generated can be interpreted with confidence for risk-related decisions.

CONTENTS

Molecular Biological Markers for Toxicology and Risk Assessment. http://dx.doi.org/10.1016/B978-0-12-809589-8.00007-X

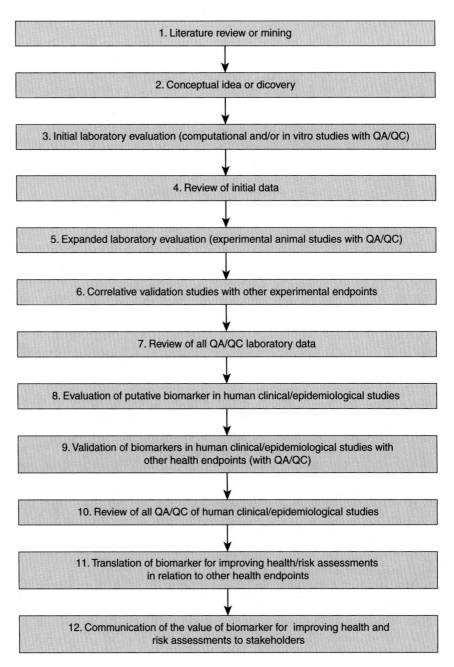

FIGURE 7.1 A general developmental flow diagram for biomarker development as an iterative biomarker used for clinical or regulatory decision making.

This figure is intended to show the process moving from conceptualization to experimental laboratory studies to clinical/epidemiological studies. Validation and QA/QC elements are considered essential at each step in the process so that the data may be used for clinical or risk assessment decision making.

2   GENERAL DEFINITIONS

2.1   Quality Assurance

In the context of molecular biomarkers for risk assessment, QA would be a formal systematic monitoring program of the fidelity of the measurements conducted and evaluation of data generated to ensure that written proscribed standards of quality are being consistently met. This aspect is particularly important when data are generated by multiple contributors, for example, at science centers or hospitals in different geographical regions or countries with plans to share samples or merge data sets.

2.2   Quality Control

For molecular biomarkers, QC is the sum total of activities designed to ensure that the standards of quality for biomarker measurements, as outlined in laboratory protocols, are being met and any errors of deviations are corrected. To achieve this goal, written QC protocols of procedures and the adherence to them for each activity are also essential.

3   DISCUSSION OF QA/QC DEFINITION

The general definitions for QA/QC noted earlier are intended to provide a framework for the various basic topic areas of QA/QC for specific molecular biomarker types (genomics, proteomics, metabolomics/metabonomics) to be discussed later in this chapter. It is clear that these procedures are important not only for the generation of good science from biomarkers but also essential for the application of these potentially powerful tools to risk assessment-based decision making. The overall purpose of implementing an effective QA/QC program is to assure data quality, reproducibility, and trust that the presented findings are correct and may be used for making clinical or regulatory decisions.

3.1   Critical Role of QA/QC in Biomarker Development and Acceptance for Risk Assessment

Biomarker development, like most fields of science, is an iterative process which involves conceptual initiation/discovery followed by various technical stages of development and testing. QA/QC is a key component of this process which usually appears at later stages of development when the biomarker is ready for application or clinical usage. A general flow chart for this process is given in Fig. 7.1.

4 SAMPLE HANDLING FOR BIOMARKER DEVELOPMENT

Handling of samples prior to any analysis is a key step in assuring generation of reliable data. This is particularly true of biological samples which are susceptible to microbial degradation. Inadequate laboratory conditions, such as inappropriate temperature for storage, photolysis for light sensitive samples (eg, porphyrins), contaminated storage containers, or inadequate storage security and sample identification can confound sample analyses prior to actual analytical measurements. Security to prevent misfiling of samples may complicate or confound biomarker data analysis. Even the most sophisticated and sensitive analytical tools cannot generate useful data if the original biological sample has been compromised by poor initial handling. This potentially serious problem is usually easy to solve with good prior planning and training of laboratory staff on laboratory protocols. The development of a written Spirit of Good Laboratory Practices (GLP) protocol l (Chen et al., 2009; Ezzelle et al., 2008) prior to beginning any biomarker development project is a good way to assure that all laboratory staff are fully informed of standard operating procedures and expectations. Similarly, new written protocols for changes in laboratory methods that result from experience or findings should be developed as molecular biomarker development evolves. The methods for conducting these analyses are also constantly being advanced. The most important overall aspect of these approaches is that there is written documentation of the procedures and adherence of handling the biomarker samples. This will enable the team or others to replicate a result if needed and be useful for instruction if there is laboratory staff turnover.

5 INTRINSIC VARIABILITY OF MEASURED BIOMARKER ENDPOINTS

As with all biological and analytical measurements, molecular biomarkers will have intrinsic variability as result of actual differences in analyte concentration in samples, sample handling techniques, degree of technician training and experience, detection limits, and instrument stability. In order to achieve statistically acceptable results, these factors must be considered and the number of samples needed to reach a satisfactory result for statistical analysis included in the overall study design through power calculations prior to initiation of the study. It is important for the proposed biomarker to be able to have sufficient capacity to distinguish effects due to an exposure/treatment from controls under the conditions of the study (eg, background exposures or small clusters of disease). This approach is of particular importance for studies whose outcome measures are to be used for the prognoses of diseases such as cancer (Cho, 2007; Hirsch et al., 2006; Laxman et al., 2008) in a clinical setting or included in MOA risk assessments which may have major regulatory implications (Borgert et al., 2004; Clewell et al., 1995).

6   EQUIPMENT MAINTENANCE, INTERNAL STANDARDS, AND CHAIN OF CUSTODY FOR ASSURANCE OF DATA QUALITY

In order to assure generation of good quality measurements for the molecular biomarkers of interest, it is important to include standard laboratory methods of QA such as regular maintenance of all analytical instruments (Ackermann et al., 2006) with documentation records, use of internal standards, chain of custody procedures, and regular testing to confirm sample identity, and confirmation of the fidelity of generated data transference for statistical analysis. Chain of custody procedures may be of particular value for data sets generated by multidisciplinary or multicenter research groups with numerous members in different geographic locations. For example, study subjects might be recruited by all participating research groups but chemical analyses conducted at only one center with special expertise and molecular biology studies at a different participating center. This would create the need to transmit and track samples so that the data could be correctly assembled at the end of the study. Clearly, tracking of data generated by such large groups may be particularly critical if the data are ultimately to be used for regulatory risk assessment purposes. Fortunately, although such multidisciplinary centers are often dispersed in different locations and present logistical challenges, they also bring needed expertise and access to larger study groups which can now be connected via data sharing on the internet. The net result is a potentially stronger science if the data from the collaborating centers is made consistent by virtue of a robust QA/QC program across the centers.

7   DATA ANALYSIS AND ARCHIVAL STORAGE NEEDS

As noted earlier, statistical analyses of molecular biomarker data are essential to reaching firm conclusions on differences in relative risk between control and exposed populations. Major advantages of molecular biomarker approaches include the sensitivity of these responses to chemical- or drug-induced perturbations which may permit delineation and adverse outcome on a relatively small number of subjects and the specificity of the response for MOA risk assessments. These types of data are potentially very valuable for making risk assessments on small groups of subjects (eg, clusters) (Anderberg, 2014) and longitudinal studies using the MOA insights to strengthen causality linkages between chronic exposures and adverse outcome effects over time. Analysis of variance and pairwise statistical methods (t-test or Mann–Whitney U test) (Birnbaum et al., 1996; Han et al., 2002; Wang et al., 2005) to compare specific groups are commonly used for such analyses. Again application of the most appropriate test will rest with the combined scientific judgment of the risk assessment professional in collaboration with a practicing statistician

(Kim et al., 2015; Mischak et al., 2015; Tournoud et al., 2015). The value of long-term biorepositories of clinical samples and appropriate QA/QC procedures is also an essential component of longitudinal studies which may require repeated sampling of biological samples for biomarker assays. Archival samples may be highly valuable for identifying important factors subsequent to the completion of the original study thereby providing information on timing. Such a program has been implemented by the eMERGE consortium (McCarty et al., 2011; Zuvich et al., 2011). This program demonstrates the value of QA/QC in permitting utilization of archived samples from multiple centers such that data from these centers may be appropriately merged for analytical studies and biomarker development. In summary, a solid QA/QC procedure program is essential to any biomarker study for the reasons outlined in this chapter.

7.1 Specific Omic QA/QC Procedures

In recent years, a number of investigators have developed QA/QC procedures for genomic, proteomic, and metabolomic/metabonomic biomarker studies. The following such procedures are available and the reader is referred to the publications cited later for specific details.

7.1.1 Genomics QA/QC

Some examples of QA/QC approaches for genomic studies are documented in the following studies (Fuscoe et al., 2007; McCarty et al., 2011; Zuvich et al., 2011). As noted earlier, the procedures and tools for conducting genomic biomarker studies are constantly evolving. For example, the advent of epigenomics require adherence to regulatory elements for biomarker studies. Appropriate QA/QC procedures should include adherence to the regulatory elements.

7.1.2 Proteomics QA/QC

The following studies contain some examples of QA/QC approaches for proteomic studies (Conrads et al., 2004; Dasari et al., 2012; Harezlak et al., 2007; Hong et al., 2005; Ma et al., 2012). As with the other omic biomarkers, the field of proteomic biomarkers is also increasing in complexity with regard to how to construct appropriate QA/QC procedures that will assure the rigor of putative proteomic biomarkers. Factors such as protein–protein interactions, posttranslational modification of proteins of interest, and noncovalent binding of oligopeptides to larger proteins which may alter their normal biological activity in vivo, all need to be considered and appropriate QA/QC procedures updated as knowledge related to QA/QC is accumulated.

7.1.3 Metabolomics/Metabonomics QA/QC

The following studies contain some useful examples of QA/QC approaches for metabolomic/metabonomic studies (Theodoridis et al., 2011; Xia et al., 2012).

Like genomic and proteomic technologies, the field of metabolomic/metabonomic biomarkers is also rapidly expanding. This means that QA/QC procedures for these types of measurements must also evolve so that the QA/QC procedures include the most current recommendations. For example, interactions between metabolic pathways may alter interpretation of the biological significance of a given metabolomic/metabonomic response.

8 SUMMARY AND CONCLUSIONS

This chapter has briefly identified some generic issues related to QA/QC needs for studies that are conducted for the development of molecular biomarkers. These issues are largely similar to those for other types of analytical measurements. They are, however, potentially more challenging to implement since the field of molecular biomarkers is rapidly evolving with new and more sensitive biomarker endpoints emerging every day. This means that there is an ongoing need for updating QA/QC approaches in order for them to be relevant and appropriate to assure good data quality for risk assessment applications.

References

Anderberg, M.R., 2014. Cluster Analysis for Applications. Probability and Mathematical Statistics: A Series of Monographs and Textbooksvol. 19Academic Press, London.

Birnbaum, D., Jacquez, G., Waller, L., Grimson, R., Wartenberg, D., 1996. The analysis of disease clusters, Part I: state of the art. Infect. Control 17 (5), 319–327.

Borgert, C.J., Quill, T.F., McCarty, L.S., Mason, A.M., 2004. Can mode of action predict mixture toxicity for risk assessment? Toxicol. Appl. Pharmacol. 201 (2), 85–96.

Burnum, K.E., Frappier, S.L., Caprioli, R.M., 2008. Matrix-assisted laser desorption/ionization imaging mass spectrometry for the investigation of proteins and peptides. Annu. Rev. Anal. Chem. 1, 689–705.

Chen, B., Gagnon, M.C., Shahangian, S., Anderson, N.L., Howerton, D.A., Boone, D.J., et al., 2009. Good laboratory practices for molecular genetic testing for heritable diseases and conditions. MMWR. 58, RR-6.

Cho, W.C., 2007. Nasopharyngeal carcinoma: molecular biomarker discovery and progress. Mol. Cancer 6 (1), 1.

Clewell, H., Gentry, P., Gearhart, J., Allen, B., Andersen, M., 1995. Considering pharmacokinetic and mechanistic information in cancer risk assessments for environmental contaminants: examples with vinyl chloride and trichloroethylene. Chemosphere 31 (1), 2561–2578.

Conrads, T.P., Fusaro, V.A., Ross, S., Johann, D., Rajapakse, V., Hitt, B.A., et al., 2004. High-resolution serum proteomic features for ovarian cancer detection. Endocr. Relat. Cancer 11 (2), 163–178.

Dasari, S., Chambers, M.C., Martinez, M.A., Carpenter, K.L., Ham, A.J., Vega-Montoto, L.J., Tabb, D.L., 2012. Pepitome: evaluating improved spectral library search for identification complementarity and quality assessment. J. Proteome Res. 11 (3), 1686–1695.

DeBord, D.G., Burgoon, L., Edwards, S.W., Haber, L.T., Kanitz, M.H., Kuempel, E., et al., 2015. Systems biology and biomarkers of early effects for occupational exposure limit setting. J. Occup. Environ. Hyg. 12, S41–S54.

Ezzelle, J., Rodriguez-Chavez, I., Darden, J., Stirewalt, M., Kunwar, N., Hitchcock, R., et al., 2008. Guidelines on good clinical laboratory practice: bridging operations between research and clinical research laboratories. J. Pharm. Biomed. Anal. 46 (1), 18–29.

Fuscoe, J.C., Tong, W., Shi, L., 2007. QA/QC issues to aid regulatory acceptance of microarray gene expression data. Environ. Mol. Mutagen. 48 (5), 349–353.

Han, W.K., Bailly, V., Abichandani, R., Thadhani, R., Bonventre, J.V., 2002. Kidney injury molecule-1 (KIM-1): a novel biomarker for human renal proximal tubule injury. Kidney Int. 62 (1), 237–244.

Harezlak, J., Wang, M., Christiani, D., Lin, X., 2007. Quantitative quality-assessment techniques to compare fractionation and depletion methods in SELDI-TOF mass spectrometry experiments. Bioinformatics 23 (18), 2441–2448.

Hirsch, F.R., Varella-Garcia, M., Bunn, P.A., Franklin, W.A., Dziadziuszko, R., Thatcher, N., et al., 2006. Molecular predictors of outcome with gefitinib in a phase III placebo-controlled study in advanced non–small-cell lung cancer. J. Clin. Oncol. 24 (31), 5034–5042.

Hong, H., Dragan, Y., Epstein, J., Teitel, C., Chen, B., Xie, Q., et al., 2005. Quality control and quality assessment of data from surface-enhanced laser desorption/ionization (SELDI) time-of flight (TOF) mass spectrometry (MS). BMC Bioinformatics, 6 Suppl. 2, S5.

Kim, E., Zeng, D., Zhou, X.H., 2015. Semiparametric transformation models for multiple continuous biomarkers in ROC analysis. Biom. J. 57, 2441–2448.

L Ackermann, B., E Hale, J., L Duffin, K., 2006. The role of mass spectrometry in biomarker discovery and measurement. Curr. Drug Metab. 7 (5), 525–539.

Laurie, C.C., Doheny, K.F., Mirel, D.B., Pugh, E.W., Bierut, L.J., Bhangale, T., et al., 2010. Quality control and quality assurance in genotypic data for genome-wide association studies. Genet. Epidemiol. 34 (6), 591–602.

Laxman, B., Morris, D.S., Yu, J., Siddiqui, J., Cao, J., Mehra, R., 2008. A first-generation multiplex biomarker analysis of urine for the early detection of prostate cancer. Cancer Res. 68 (3), 645–649.

Ma, Z.Q., Polzin, K.O., Dasari, S., Chambers, M.C., Schilling, B., Gibson, B.W., et al., 2012. Qua-Meter: multivendor performance metrics for LC-MS/MS proteomics instrumentation. Anal. Chem. 84 (14), 5845–5850.

McCarty, C.A., Chisholm, R.L., Chute, C.G., Kullo, I.J., Jarvik, G.P., Larson, E.B., et al., 2011. The eMERGE Network: a consortium of biorepositories linked to electronic medical records data for conducting genomic studies. BMC Med. Genomics 4, 13.

Mischak, H., Critselis, E., Hanash, S., Gallagher, W.M., Vlahou, A., Ioannidis, J.P., 2015. Epidemiologic design and analysis for proteomic studies: a primer on -omic technologies. Am. J. Epidemiol. 181 (9), 635–647.

Theodoridis, G., Gika, H.G., Wilson, I.D., 2011. Mass spectrometry-based holistic analytical approaches for metabolite profiling in systems biology studies. Mass Spectrom. Rev. 30 (5), 884–906.

Tournoud, M., Larue, A., Cazalis, M.A., Venet, F., Pachot, A., Monneret, G., et al., 2015. A strategy to build and validate a prognostic biomarker model based on RT-qPCR gene expression and clinical covariates. BMC Bioinformatics 16, 106.

Wagner, G., 1995. Basic approaches and methods for quality assurance and quality control in sample collection and storage for environmental monitoring. Sci. Total Environ. 176 (1–3), 63–71.

Wang, C., Kong, H., Guan, Y., Yang, J., Gu, J., Yang, S., Xu, G., 2005. Plasma phospholipid metabolic profiling and biomarkers of type 2 diabetes mellitus based on high-performance liquid chromatography/electrospray mass spectrometry and multivariate statistical analysis. Anal. Chem. 77 (13), 4108–4116.

Whitney, C.W., Lind, B.K., Wahl, P.W., 1998. Quality assurance and quality control in longitudinal studies. Epidemiol. Rev. 20 (1), 71–80.

Xia, J., Mandal, R., Sinelnikov, I.V., Broadhurst, D., Wishart, D.S., 2012. MetaboAnalyst 2.0—a comprehensive server for metabolomic data analysis. Nucleic Acids Res. 40 (Web Server issue), W127–133.

Zuvich, R.L., Armstrong, L.L., Bielinski, S.J., Bradford, Y., Carlson, C.S., Crawford, D.C., et al., 2011. Pitfalls of merging GWAS data: lessons learned in the eMERGE network and quality control procedures to maintain high data quality. Genet. Epidemiol. 35 (8), 887–898.

CHAPTER 8

Translation of Biomarkers for Human Clinical and Epidemiological Studies

1 INTRODUCTION

As noted in previous chapters of this book, molecular biomarkers may be classified by the type of biological response they reflect (eg, genomic, proteomic, metabolomic/metabonomic, etc.). Biomarkers may also be viewed by the degree of the biological response they represent such as biomarkers of exposure, biomarkers of moderate effect, biomarkers cell injury/cell death, cancer biomarkers, and biomarkers of birth defects which manifest themselves at birth or at a later life stage such as diabetes (Alonso-Magdalena et al., 2011; Lee et al., 2009; Newbold et al., 2007). In addition, the direct impact of chemical/drug exposures on persons in childhood due to developing organ systems are well appreciated as major and growing public health problems (Ginsberg et al., 2005). Older persons are generally more sensitive to such chemical/drug exposures for this reason and hence become a sensitive subpopulation, at special risk, for toxicity (Fowler, 2009, 2012, 2013).

In all these cases, the value of a given biomarker endpoint for risk assessment rests with its prognostic significance and reliability. This chapter will briefly examine some of the currently more well-developed types of molecular biomarkers in the previously mentioned areas from the perspective of the level of biological response they have been documented to reflect. It must be remembered, as noted elsewhere in this book, that the field of molecular biomarkers is moving forward at an ever-increasing pace and that new classes of biomarkers will undoubtedly emerge in the near future so it is not possible to be inclusive of all molecular biomarkers. This chapter will attempt to provide an overview of *some of the ways* the current major types of biomarker classes may be functionally linked to major general biological response pathways and suggest some possible ways this mode of action (MOA) information could be further translated for practical risk assessment purposes and adverse outcome pathways (AOPs) approaches to provide improved protection of public health. It is not possible here to provide an all-encompassing review of the literature on each of the biomarkers since, as previously noted in this book, the literature

CONTENTS

Molecular Biological Markers for Toxicology and Risk Assessment. http://dx.doi.org/10.1016/B978-0-12-809589-8.00008-1

on biomarkers is already large and growing rapidly. For this reason, this chapter will attempt to review a number of the more current biomarker classes and provide key references for further reading.

2 CLEAR DEFINITIONS OF BIOMARKER TERMINOLOGY

It is important to start this discussion by outlining some general definitions of biomarker terminology. The omic biomarker classes have been extensively discussed in previous chapters and hence a description of them will not be repeated here. For purposes of this discussion, biomarkers may also be classified by where they have been shown to provide useful information on adverse risk effects from a documented chemical/drug exposure to major types of overt clinical disease, cell death, birth defects, or cancer. This chapter will focus on how biomarkers may be used functionally to help detect and elucidate early cellular responses to chemicals and assist in the prevention of overt organ system damage or cancer.

2.1 Biomarkers of Exposure

2.1.1 Biomonitoring for Chemical/Drug Exposures

Biomonitoring programs such as the NHANES (CDC, 2005) operated by CDC is a very valuable program which utilizes highly sophisticated computer-operated analytical equipment to measure concentrations of drugs and chemicals in human tissues and fluids to assess drug/chemical exposures. The program directly measures chemicals and metabolites which are used as biomarkers of exposure.

2.1.2 Biology-Based Biomarkers of Chemical/Drug Exposures

Biology-based biomarkers are among the earliest responses to chemical exposures, which show responses to chemical exposures prior to or earlier than more overt disturbances in major cellular metabolic systems. The binding of lead and inhibition of the heme biosynthetic pathway enzyme ALAD (Goering and Fowler, 1984, 1985) would be an example of the former and induction of specific stress protein families by the III–V semiconductors GaAs and InAs (Fowler et al., 2005, 2008) would be examples of the latter. The main point here is that chemical interactions with these very sensitive molecules will produce measureable chemical and/or mixture-specific measureable responses which could be used for MOA risk assessments.

2.2 Biomarkers of Effects

Biomarkers of more overt cellular effects are generally linked to the physical disorganization of normal cellular metabolic machinery such as the mitochondria (Fowler and Woods, 1979) which is associated with disruption of a number of essential metabolic pathways such as those producing energy

(Fowler et al., 1979) and heme to meet cellular metabolic system needs. The disruption of these essential pathways leads to generation of ROS from inhibition of cellular respiration (Inoue et al., 2003; Liu et al., 2002; Weinberg et al., 2010) which is further exacerbated by the accumulation of ALA and porphyrins that are also capable of catalyzing ROS formation (Fuchs et al., 2000; O'Connor et al., 2009). These changes are often among the earliest disruptive effects produced by a number drugs and chemical which antecede the onset of overt cell death (Higuchi et al., 1998), apoptosis (Efferth et al., 2007; Higuchi et al., 1998; Vrablic et al., 2001), DNA damage (Ding et al., 2005), and cancer (Weinberg et al., 2010). It should be noted here that biomarkers of effects may be reversible due to cessation of chemical/drug exposure or the action of cellular defense mechanisms (Fowler et al., 2008; Whittaker et al., 2011).

2.2.1 Biomarkers of Toxicity

In recent years, extensive attention has been given to the biomarkers of early cellular toxicity which are part of a continuum with biomarkers of early effects and the biomarkers of cell injury and cell death (Please see Chapter 5 for further discussion). There are potential classes of biomarkers of toxicity linked to early manifestations of toxicity. Ubiquitin tagging of proteins prior to degradation by the proteasome (Ciechanover, 1998; Finley, 2009; Lecker et al., 2006; Verma et al., 2000), increased phagocytosis of organelles with lysosomal degradation (Klionsky and Emr, 2000; Mizushima et al., 2008; Yoshimori, 2004), and cellular swelling (Danial and Korsmeyer, 2004; Trump et al., 1997) due to membrane damage and activation of the FAS receptor system (Galle et al., 1995; Matsumura et al., 2000; Vercammen et al., 1998) are generally regarded as further steps along the road to initiation of cell death (Danial and Korsmeyer, 2004; Lemasters et al., 1998; Majno and Joris, 1995). There is sufficient basic scientific literature on these processes to consider how this information could be translated for the development of biomarkers to support MOA risk assessment activities (Andersen and Dennison, 2001; Borgert et al., 2004) since there are antibodies and other molecular tools currently available to follow the activities of these important cellular systems.

2.2.2 Biomarkers of Cell Injury/Cell Death

Apoptosis and necrosis are the major general processes by which cells die and have received extensive scientific attention over the last several decades (Alderson et al., 1995; Iwai et al., 1994; Obeid et al., 2007; Orrenius et al., 2003; Suzuki et al., 2000). There is a great deal of basic scientific information on these two processes and their relationships to each other and to provide clearer definitions of the AOPs discussed later. As discussed earlier, there is a lot of good molecular information which could also be mined and used for the development of biomarker assays (eg, immune assays) which support MOA risk assessments by documenting initiated events in cell death processes.

2.2.3 Adverse Outcome Pathways

AOPs represent a risk assessment framework which is intended to link initiating toxic events at the molecular level such as those discussed earlier with the risk assessment process (Jennings, 2013; Schultz, 2010; Vinken, 2013). In terms of how biomarkers may be helpful, it is hence a conceptual approach by which molecular biomarker responses may be translated to help support risk assessment conclusions by providing a basic scientific foundation for these judgments. AOPs are hence potentially valuable as a means by which biomarkers may be accepted as valuable integral components of modern risk assessment practice. Fig. 8.1 is a general diagram which shows how the earlier sections relate to each other and feed into supporting AOPs as the conceptual portal by which basic scientific molecular biomarker information may be used to support risk assessment conclusions.

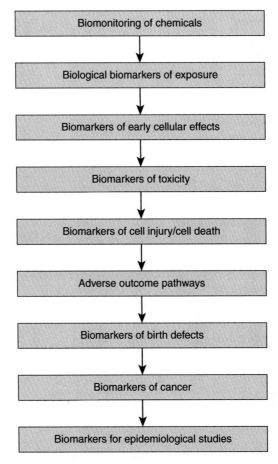

FIGURE 8.1 A short list of the major types of biomonitoring, exposure, and adverse health outcome categories where biomarkers have provided information of value to risk assessments.

2.3 Biomarkers of Cancer and Birth Defects

The concepts discussed earlier in relation to development of AOPs for risk assessment largely assume chemical exposures to mature and stable cell populations in target organs. The issue of how to apply biomarkers information to rapidly dividing cell populations such as cancer cells or developing organ systems is a special case which deserves separate attention since these cells express a number of basic biological differences which also impact the usual or expected biomarker responses. These differences mean that interpretation of biomarker responses must also take the biology of cancer or developing cell types into account. These cells are intrinsically different from normal cell populations and those differences must be considered in utilizing biomarker data derived from such cells for developing AOP-based risk assessment conclusions. For example, high amplitude swelling of mitochondria from adult animals' exposure to agents such as methyl mercury is a common biological response associated with a number of biochemical disturbances (Fowler et al., 1975; Fowler and Woods, 1977b). Similar exposures to fetal animals in utero produce a reduction of mitochondrial volume density and suppression of a number of metabolic functions related to inhibition of mitochondrial biogenesis in the developing organism (Fowler and Woods, 1977a). These disturbances persist into later life stages (Fowler, 1982). The point here is that the biological response at the level of the mitochondrion is different depending upon the stage in life when the exposure occurs.

Altered cellular imprinting in the developing embryo from in utero exposure to agents such as methyl mercury or persistent organic pollutants such as dioxins, PCBs, or DDT (Desaulniers et al., 2009; Lambertini et al., 2012) may lead to major adverse health consequences later in life. Basic scientific knowledge of these metabolic pathway disruptions could lead to the measurement of key regulatory proteins in these processes as biomarkers of effects for the identified health outcomes.

3 BIOMARKERS FOR EPIDEMIOLOGICAL STUDIES

3.1 Discussion of Strengths and Weaknesses of Biomarkers for Epidemiology Studies

It is clear from the previous discussion that molecular biomarkers when fully validated and translated for risk assessment practice could become vital tools in the field of epidemiology for identifying and explaining associations between drug or chemical exposures and a variety of disease endpoints. Providing MOA-based information on the processes underlying the development of clinical diseases resulting from chronic low dose exposures to a variety of chemical agents on an individual or mixture basis is a major problem in both longitudinal and cluster epidemiology studies as discussed later. The incorporation of molecular biomarkers into these types of studies has the potential for early detection of individuals at special risk for adverse health outcomes in these types of studies but also adding the element of prevention of disease for this reason.

3.1.1 Longitudinal Studies

Longitudinal epidemiology studies traditionally follow large cohorts of individuals through time and observe the incidence of various disease endpoints in relation to drug or chemical exposures. This is a useful information for establishing causal associations using the Bradford Hill Criteria (Hill, 1965) discussed elsewhere in this book. These types of studies would be more valuable if they were able to provide mechanistic information on the underlying chemical-induced disruptions leading to the development of the disease endpoints and identification of those individuals in the cohort at special risk. Inclusion of molecular biomarkers into such studies would hence provide a means for both explaining why a particular disease developed in the study population on a mechanistic basis and offer the opportunity for prevention of further disease development by reducing or stopping the causative chemical exposures.

3.1.2 Cluster Studies

Studies of clusters of diseases are problematic in epidemiology due to usually small numbers of individual involved. Unless the disease endpoint is relatively rare such as mesothelioma (Boffetta, 2007) in relation to asbestos exposure or hemangiosarcoma in relation to vinyl chloride (Ward et al., 2001) exposures, linking common types of cancers or chronic diseases such as diabetes to chemical exposures is usually statistically difficult due to the small number of individuals included in the study population and the relatively large numbers of persons in the general population expressing the disease. Inclusion of validated molecular biomarkers capable of showing 10-fold differences in response to a specific chemical exposure (Woods and Fowler, 1977, 1978) would greatly increase the sensitivity of cluster studies to reach firm conclusions and permit MOA-based risk assessment decisions.

3.2 Birth/Metabolic Defect Studies From In utero Exposures

It is increasingly clear that in utero exposures to drugs (eg, thalidomide) or chemicals (BPA) can produce a spectrum of effects in offspring which may manifest themselves at birth as overt physical or more subtle birth defects (Heindel, 2005; ten Tusscher et al., 2000) or later on in life as metabolic diseases such as type 2 diabetes and insulin resistance (Lee et al., 2006, 2007). Major problems in this area are: (1) prevention of these disorders with early detection of causative processes and (2) estimating risk of these effects via disturbances in AOPs. On a mechanistic level, many of the pathways involved in regulation and control of embryogenesis have been well-studied (Chugh and Khurana, 2002; Hutvagner and Simard, 2008) and assays for key proteins in these pathways could be turned into molecular biomarkers via development of immunoassay kits to look for early signs of specific perturbations in response to drug or chemical exposures.

4 BIOMARKERS FOR CHEMICAL MIXTURE RISK ASSESSMENTS

Chemical mixtures are another problematic area of increasing public health concern since interactions between agents in a mixture may greatly alter the shape of an expected dose-response curve in an additive, synergistic, or antagonistic manner. Such interactions may hence strongly influence the accuracy of risk assessment conclusions and predictions. Development of molecular biomarkers for key regulatory proteins or biochemical processes/metabolic products in AOPs which are responsive at low dose levels to one or more components in a mixture of drugs, chemicals, or drugs plus chemicals would be highly valuable for improving the quality of risk assessments. One example of such an approach is following interactions between lead, cadmium, and arsenic in food (Mahaffey et al., 1981) or drinking water (Whittaker et al., 2011) via disturbances in the heme biosynthetic pathway. These studies, conducted in rodents, demonstrated additive interactions in these common mixtures at both intermediate "stressor" or low dose levels, respectively.

5 MERGING CHEMICAL EXPOSURE DATA AND GENETIC INHERITANCE DATA FOR RISK ASSESSMENTS

In addition to dose and duration of exposure (eg, dose-response/time-course), there is increasing evidence of the importance of gender (Fowler, 2012; Fowler et al., 2008), age at exposure (Ruiz et al., 2010), and genetic inheritance in mediating susceptibility to drug or chemical toxicities (Chang et al., 2009; Scinicariello et al., 2005, 2010). Molecular biomarkers are extremely valuable for providing MOA information on subpopulations at special risk for toxicity. This type of information can help identify, at the molecular level, why such individuals are sensitive and hence help focus supplemental data studies into productive channels.

6 SUMMARY AND CONCLUSIONS

This chapter has attempted to examine molecular biomarkers from the perspective of the types of biological processes or phenomena they attempt to measure and to suggest ways in which information derived from biomarkers could be translated to strengthen risk assessment practice. It is also clear that the various biomarkers discussed are in different stages of development based upon current scientific understandings of their biological significance. This is a central issue in the process of translating putative biomarkers for risk assessment purposes (eg, what does a change in a validated biomarker mean in biological terms and hence what is its prognostic significance?). While a small number of currently developed biomarkers are presently understood to this level, much remains to be done if biomarkers are to reach their full potential and acceptance as cornerstones in modern MOA-based risk assessment practice. The public health benefits of having molecular biomarker tools incorporated into risk assessment practice to help address some of the pressing problems discussed earlier are potentially profound.

References

Alderson, M.R., Tough, T.W., Davis-Smith, T., Braddy, S., Falk, B., Schooley, K.A., et al.,1995. Fas ligand mediates activation-induced cell death in human T lymphocytes. J. Exp. Med. 181 (1), 71–77.

Alonso-Magdalena, P., Quesada, I., Nadal, A., 2011. Endocrine disruptors in the etiology of type 2 diabetes mellitus. Nat. Rev. Endocrinol. 7 (6), 346–353.

Andersen, M.E., Dennison, J.E., 2001. Mode of action and tissue dosimetry in current and future risk assessments. Sci. Total Environ. 274 (1), 3–14.

Boffetta, P., 2007. Epidemiology of peritoneal mesothelioma: a review. Ann. Oncol. 18 (6), 985–990.

Borgert, C.J., Quill, T.F., McCarty, L.S., Mason, A.M., 2004. Can mode of action predict mixture toxicity for risk assessment? Toxicol. Appl. Pharmacol. 201 (2), 85–96.

CDC, 2005. Third national report on human exposure to environmental chemicals. National Center for Environmental Health, NCEH Pub. 05-0570.

Chang, M.H., Lindegren, M.L., Butler, M.A., Chanock, S.J., Dowling, N.F., Gallagher, M., et al.,2009. Prevalence in the United States of selected candidate gene variants: Third National Health and Nutrition Examination Survey, 1991–1994. Am. J. Epidemiol. 169 (1), 54–66.

Chugh, A., Khurana, P., 2002. Gene expression during somatic embryogenesis-recent advances. Curr. Sci. 83 (6), 715–730.

Ciechanover, A., 1998. The ubiquitin–proteasome pathway: on protein death and cell life. EMBO J. 17 (24), 7151–7160.

Danial, N.N., Korsmeyer, S.J., 2004. Cell death: critical control points. Cell 116 (2), 205–219.

Desaulniers, D., Xiao, G.-H., Lian, H., Feng, Y.-L., Zhu, J., Nakai, J., Bowers, W.J., 2009. Effects of mixtures of polychlorinated biphenyls, methylmercury, and organochlorine pesticides on hepatic DNA methylation in prepubertal female Sprague–Dawley rats. Int. J. Toxicol. 28 (4), 294–307.

Ding, W., Hudson, L.G., Liu, K.J., 2005. Inorganic arsenic compounds cause oxidative damage to DNA and protein by inducing ROS and RNS generation in human keratinocytes. Mol. Cell. Biochem. 279 (1–2), 105–112.

Efferth, T., Giaisi, M., Merling, A., Krammer, P.H., Li-Weber, M., 2007. Artesunate induces ROS-mediated apoptosis in doxorubicin-resistant T leukemia cells. PLoS One 2 (8), e693.

Finley, D., 2009. Recognition and processing of ubiquitin-protein conjugates by the proteasome. Annu. Rev. Biochem. 78, 477.

Fowler, B.A., 1982. Ultrastructural and biochemical alterations of cellular organelle systems by prenatal exposure to toxic trace metals. Proceedings of International Conference on the Developmental and Reproductive Toxicity of Metals. Plenum Press, University of Rochester, Rochester, New York, May 1982. 1983 pp.

Fowler, B.A., 2009. Monitoring of human populations for early markers of cadmium toxicity: a review. Toxicol. Appl. Pharmacol. 238 (3), 294–300.

Fowler, B.A., 2012. Biomarkers in toxicology and risk assessment. EXS 101, 459–470.

Fowler, B.A., 2013. Cadmium and aging. In: Weiss, B. (Ed.), Aging and Vulnerabilities to Environmental Chemicals. Royal Society of Chemistry, Cambridge, UK, pp. 376–387.

Fowler, B.A., Woods, J.S., 1977a. The transplacental toxicity of methyl mercury to fetal rat liver mitochondria. Morphometric and biochemical studies. Lab. Invest. 36 (2), 122–130.

Fowler, B.A., Woods, J.S., 1977b. Ultrastructural and biochemical changes in renal mitochondria during chronic oral methyl mercury exposure: the relationship to renal function. Exp. Mol. Pathol. 27 (3), 403–412.

Fowler, B.A., Woods, J.S., 1979. The effects of prolonged oral arsenate exposure on liver mitochondria of mice: morphometric and biochemical studies. Toxicol. Appl. Pharmacol. 50 (2), 177–187.

Fowler, B.A., Brown, H.W., Lucier, G.W., Krigman, M.R., 1975. The effects of chronic oral methyl mercury exposure on the lysosome system of rat kidney. Morphometric and biochemical studies. Lab. Invest. 32 (3), 313–322.

Fowler, B.A., Woods, J.S., Schiller, C.M., 1979. Studies of hepatic mitochondrial structure and function: morphometric and biochemical evaluation of in vivo perturbation by arsenate. Lab. Invest. 41 (4), 313–320.

Fowler, B.A., Conner, E.A., Yamauchi, H., 2005. Metabolomic and proteomic biomarkers for III-V semiconductors: chemical-specific porphyrinurias and proteinurias. Toxicol. Appl. Pharmacol. 206 (2), 121–130.

Fowler, B.A., Conner, E.A., Yamauchi, H., 2008. Proteomic and metabolomic biomarkers for III-V semiconductors: and prospects for application to nano-materials. Toxicol. Appl. Pharmacol. 233 (1), 110–115.

Fuchs, J., Weber, S., Kaufmann, R., 2000. Genotoxic potential of porphyrin type photosensitizers with particular emphasis on 5-aminolevulinic acid: implications for clinical photodynamic therapy. Free Radic. Biol. Med. 28 (4), 537–548.

Galle, P.R., Hofmann, W.J., Walczak, H., Schaller, H., Otto, G., Stremmel, W., et al.,1995. Involvement of the CD95 (APO-1/Fas) receptor and ligand in liver damage. J. Exp. Med. 182 (5), 1223–1230.

Ginsberg, G., Hattis, D., Russ, A., Sonawane, B., 2005. Pharmacokinetic and pharmacodynamic factors that can affect sensitivity to neurotoxic sequelae in elderly individuals. Environ. Health Perspect. 113, 1243–1249.

Goering, P.L., Fowler, B.A., 1984. Regulation of lead inhibition of delta-aminolevulinic acid dehydratase by a low molecular weight, high affinity renal lead-binding protein. J. Pharmacol. Exp. Ther. 231 (1), 66–71.

Goering, P.L., Fowler, B.A., 1985. Mechanism of renal lead-binding protein reversal of delta-aminolevulinic acid dehydratase inhibition by lead. J. Pharmacol. Exp. Ther. 234 (2), 365–371.

Heindel, J.J., 2005. The fetal basis of adult disease: role of environmental exposures—introduction. Birth Defects Res. A 73 (3), 131–132.

Higuchi, M., Honda, T., Proske, R.J., Yeh, E.T., 1998. Regulation of reactive oxygen species-induced apoptosis and necrosis by caspase 3-like proteases. Oncogene 17 (21), 2753–2760.

Hill, A.B., 1965. The environment and disease: association or causation? Proc. Royal Soc. Med. 58 (5), 295–300.

Hutvagner, G., Simard, M.J., 2008. Argonaute proteins: key players in RNA silencing. Nat. Rev. Mol. Cell Biol. 9 (1), 22–32.

Inoue, M., Sato, E.F., Nishikawa, M., Park, A.-M., Kira, Y., Imada, I., Utsumi, K., 2003. Mitochondrial generation of reactive oxygen species and its role in aerobic life. Curr. Med. Chem. 10 (23), 2495–2505.

Iwai, K., Miyawaki, T., Takizawa, T., Konno, A., Ohta, K., Yachie, A., et al., 1994. Differential expression of bcl-2 and susceptibility to anti-Fas-mediated cell death in peripheral blood lymphocytes, monocytes, and neutrophils. Blood 84 (4), 1201–1208.

Jennings, P., 2013. Stress response pathways, toxicity pathways and adverse outcome pathways. Arch. Toxicol. 87 (1), 13–14.

Klionsky, D.J., Emr, S.D., 2000. Autophagy as a regulated pathway of cellular degradation. Science 290 (5497), 1717–1721.

Lambertini, L., Marsit, C.J., Chen, J., Lee, M.-J., 2012. Genomic imprinting in human placenta. Intech Open Access Publisher, Rijeka, Croatia.

Lecker, S.H., Goldberg, A.L., Mitch, W.E., 2006. Protein degradation by the ubiquitin–proteasome pathway in normal and disease states. J. Am. Soc. Nephrol. 17 (7), 1807–1819.

Lee, D.H., Lee, I.K., Song, K., Steffes, M., Toscano, W., Baker, B.A., Jacobs, Jr., D.R., 2006. A strong dose-response relation between serum concentrations of persistent organic pollutants and diabetes: results from the National Health and Examination Survey 1999–2002. Diabetes Care 29 (7), 1638–1644.

Lee, D.H., Lee, I.K., Jin, S.H., Steffes, M., Jacobs, Jr., D.R., 2007. Association between serum concentrations of persistent organic pollutants and insulin resistance among nondiabetic adults: results from the National Health and Nutrition Examination Survey 1999–2002. Diabetes Care 30 (3), 622–628.

Lee, D.-H., Jacobs, Jr., D.R., Porta, M., 2009. Hypothesis: a unifying mechanism for nutrition and chemicals as lifelong modulators of DNA hypomethylation. Environ. Health Perspect. 117, 1799–1802.

Lemasters, J.J., Nieminen, A.-L., Qian, T., Trost, L.C., Elmore, S.P., Nishimura, Y., et al., 1998. The mitochondrial permeability transition in cell death: a common mechanism in necrosis, apoptosis and autophagy. Biochim. Biophys. Acta 1366 (1), 177–196.

Liu, Y., Fiskum, G., Schubert, D., 2002. Generation of reactive oxygen species by the mitochondrial electron transport chain. J. Neurochem. 80 (5), 780–787.

Mahaffey, K.R., Capar, S.G., Gladen, B.C., Fowler, B.A., 1981. Concurrent exposure to lead, cadmium, and arsenic. Effects on toxicity and tissue metal concentrations in the rat. J. Lab. Clin. Med. 98 (4), 463–481.

Majno, G., Joris, I., 1995. Apoptosis, oncosis, and necrosis. An overview of cell death. Am. J. Pathol. 146 (1), 3.

Matsumura, H., Shimizu, Y., Ohsawa, Y., Kawahara, A., Uchiyama, Y., Nagata, S., 2000. Necrotic death pathway in Fas receptor signaling. J. Cell Biol. 151 (6), 1247–1256.

Mizushima, N., Levine, B., Cuervo, A.M., Klionsky, D.J., 2008. Autophagy fights disease through cellular self-digestion. Nature 451 (7182), 1069–1075.

Newbold, R.R., Padilla-Banks, E., Snyder, R.J., Phillips, T.M., Jefferson, W.N., 2007. Developmental exposure to endocrine disruptors and the obesity epidemic. Reprod. Toxicol. 23 (3), 290–296.

O'Connor, A.E., Gallagher, W.M., Byrne, A.T., 2009. Porphyrin and nonporphyrin photosensitizers in oncology: preclinical and clinical advances in photodynamic therapy. Photochem. Photobiol. 85 (5), 1053–1074.

Obeid, M., Tesniere, A., Ghiringhelli, F., Fimia, G.M., Apetoh, L., Perfettini, J.-L., et al., 2007. Calreticulin exposure dictates the immunogenicity of cancer cell death. Nat. Med. 13 (1), 54–61.

Orrenius, S., Zhivotovsky, B., Nicotera, P., 2003. Regulation of cell death: the calcium-apoptosis link. Nat. Rev. Mol. Cell Biol. 4 (7), 552–565.

Ruiz, P., Mumtaz, M., Osterloh, J., Fisher, J., Fowler, B.A., 2010. Interpreting NHANES biomonitoring data, cadmium. Toxicol. Lett. 198 (1), 44–48.

Schultz, T., 2010. Adverse outcome pathways: a way of linking chemical structure to in vivo toxicological hazards. In Silico Toxicology: Principles and Applications, Royal Society of Chemistry, Cambridge, UK, pp. 346–371.

Scinicariello, F., Murray, H.E., Smith, L., Wilbur, S., Fowler, B.A., 2005. Genetic factors that might lead to different responses in individuals exposed to perchlorate. Environ. Health Perspect. 113 (11), 1479–1484.

Scinicariello, F., Yesupriya, A., Chang, M.H., Fowler, B.A., 2010. Modification by ALAD of the association between blood lead and blood pressure in the U.S. population: results from the Third National Health and Nutrition Examination Survey. Environ. Health Perspect. 118 (2), 259–264.

Suzuki, A., Ito, T., Kawano, H., Hayashida, M., Hayasaki, Y., Tsutomi, Y., et al.,2000. Survivin initiates procaspase 3/p21 complex formation as a result of interaction with Cdk4 to resist Fas-mediated cell death. Oncogene 19 (10), 1346–1353.

ten Tusscher, G.W., Stam, G.A., Koppe, J.G., 2000. Open chemical combustions resulting in a local increased incidence of orofacial clefts. Chemosphere 40 (9), 1263–1270.

Trump, B.E., Berezesky, I.K., Chang, S.H., Phelps, P.C., 1997. The pathways of cell death: oncosis, apoptosis, and necrosis. Toxicol. Pathol. 25 (1), 82–88.

Vercammen, D., Brouckaert, G., Denecker, G., Van de Craen, M., Declercq, W., Fiers, W., Vandenabeele, P., 1998. Dual signaling of the Fas receptor: initiation of both apoptotic and necrotic cell death pathways. J. Exp. Med. 188 (5), 919–930.

Verma, R., Chen, S., Feldman, R., Schieltz, D., Yates, J., Dohmen, J., Deshaies, R.J., 2000. Proteasomal proteomics: identification of nucleotide-sensitive proteasome-interacting proteins by mass spectrometric analysis of affinity-purified proteasomes. Mol. Biol. Cell 11 (10), 3425–3439.

Vinken, M., 2013. The adverse outcome pathway concept: a pragmatic tool in toxicology. Toxicology 312, 158–165.

Vrablic, A.S., Albright, C.D., Craciunescu, C.N., Salganik, R.I., Zeisel, S.H., 2001. Altered mitochondrial function and overgeneration of reactive oxygen species precede the induction of apoptosis by 1-O-octadecyl-2-methyl-rac-glycero-3-phosphocholine in p53-defective hepatocytes. FASEB J. 15 (10), 1739–1744.

Ward, E., Boffetta, P., Andersen, A., Colin, D., Comba, P., Deddens, J.A., et al.,2001. Update of the follow-up of mortality and cancer incidence among European workers employed in the vinyl chloride industry. Epidemiology 12 (6), 710–718.

Weinberg, F., Hamanaka, R., Wheaton, W.W., Weinberg, S., Joseph, J., Lopez, M., et al.,2010. Mitochondrial metabolism and ROS generation are essential for Kras-mediated tumorigenicity. Proc. Natl. Acad. Sci. 107 (19), 8788–8793.

Whittaker, M.H., Wang, G., Chen, X.Q., Lipsky, M., Smith, D., Gwiazda, R., Fowler, B.A., 2011. Exposure to Pb, Cd, and As mixtures potentiates the production of oxidative stress precursors: 30-day, 90-day, and 180-day drinking water studies in rats. Toxicol. Appl. Pharmacol. 254 (2), 154–166.

Woods, J.S., Fowler, B.A., 1977. Renal porphyrinuria during chronic methyl mercury exposure. J. Lab. Clin. Med. 90 (2), 266–272.

Woods, J.S., Fowler, B.A., 1978. Altered regulation of mammalian hepatic heme biosynthesis and urinary porphyrin excretion during prolonged exposure to sodium arsenate. Toxicol. Appl. Pharmacol. 43 (2), 361–371.

Yoshimori, T., 2004. Autophagy: a regulated bulk degradation process inside cells. Biochem. Biophys. Res. Commun. 313 (2), 453–458.

CHAPTER 9

Risk Communication of Molecular Biomarker Information

1 INTRODUCTION

1.1 Translation of Biomarker Data Into "Plain English"

In order for molecular biomarkers to again broad acceptance and use in clinical practice and risk assessment, it is important that highly technical and rapidly evolving molecular biology information be translated into language which is more readily understood by persons with less technical backgrounds for example, "plain English" (Cutts, 2013; Thrush, 2001). This goal can be achieved with some modest effort but is worth the trouble for several reasons. First, it is most important that persons receiving the information (lay public, patients, regulators, legislators) have a clear understanding of the significance of the data in order to appreciate and act on it. Many of the terms in molecular biology are essentially similar to a highly technical foreign language which is rapidly evolving. Practitioners in the field need to be aware of this consideration and make an effort to communicate findings in readily understandable terms. Techniques such as information mapping technology as discussed later may be useful in this regard. Other practical reasons for making this effort at communication is recognition of the fact that many people obtain information from the Internet and social media which are usually limited in permitted degree of detail. Translation of molecular biomarkers information into languages other than English is also an important consideration in order to reach the global community. Many foreign languages are less flexible than English and it is easier to translate information which has already been put in plain English in order to communicate more efficiently and have meanings accurately understood.

1.2 Biomarker Tutorials for Managers and Societal Decision Makers

In terms of moving the field of molecular biomarkers forward and gaining support for future development, it is essential that managers and societal decision makers understand both the value of molecular biomarkers in meeting risk assessment decisions and also where the field is going in terms of future

CONTENTS

Molecular Biological Markers for Toxicology and Risk Assessment. http://dx.doi.org/10.1016/B978-0-12-809589-8.00009-3

developments. One approach for addressing these needs is periodic tutorials provided as either in person lectures or CE training courses at meetings, agencies, places of business, or as webinars over the Internet to reach larger audiences (Mohorovičić et al., 2011; Molay, 2009; Zaragoza-Anderson, 2008). These venues are currently being used to good effect for a variety of teaching and promotional activities (Guanci, 2009; McCarthy et al., 2012) so there is a good track record in place and no reason why training information on molecular biomarkers could not be handled in a similar manner.

2   INFORMATION MAPPING TECHNOLOGY

Information mapping technology is an approach to teaching which is commonly used in modern textbooks to increase understanding and retention of information by students (Ellis et al., 1993; Horn, 1974; Houghton et al., 1996). The technique basically involves presenting a concept or idea in clear terms as a simple question and answering the question in adjacent text in a succinct and clear manner. Additional reference information and web links are given at the end of the presentation for further reading. The overall approach is to provide the student with the most central conceptual information and more detailed background reference information in a compact readable package. A draft example of an information mapping page describing application of a metabolomic biomarker (arsenic-induced porphyrinuria pattern) to an MOA arsenic risk assessment is provided below using arsenical disruption of the heme biosynthetic pathway and attendant porphyrinuria pattern as an example. This is a brief example of how the information mapping technique may be used to translate complex biochemical information on the mechanisms of arsenical-induced disruption of the heme biosynthetic pathway leading to an arsenic-specific porphyrinuria pattern (Woods and Fowler, 1978) which could be applied to support a MOA-based risk assessment in humans exposed to arsenic in drinking water (Garcia-Vargas et al., 1994; García-Vargas and Hernández-Zavala, 1996; Hernández-Zavala et al., 1999) and from burning of coal (Deng et al., 2007; Wang et al., 2002; Xie et al., 2001). The following discussion will focus on the potential use of arsenical disruption of the heme biosynthetic pathway and attendant characteristic porphyrinuria pattern for developing an MOA-based risk assessment in humans exposed to arsenic in drinking water.

1. *What is the heme biosynthetic pathway?*

    The heme pathway is essential for life and is highly conserved across species for the production of heme which is used in a host of oxidation/reduction reactions to produce cellular energy and metabolism of both internal and external chemicals. This pathway is also sensitive to perturbation by a number of organic an inorganic chemicals at various steps resulting in specific porphyrinuria patterns. The heme biosynthetic pathway consists of enzymes located in both the mitochondria and cytoplasm of the cell (Fig. 9.1) The heme degradative side of the

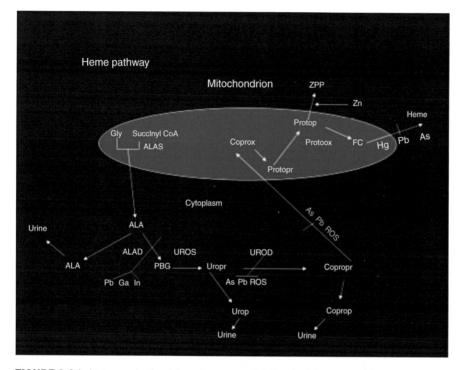

FIGURE 9.1 A short example of an information mapped briefing sheet for an moa risk assessment based on known arsenic-induced disturbances of the heme biosynthetic pathway.
A global diagram of the heme biosynthetic pathway showing the enzymatic steps and known places where exposure to toxic elements, including arsenic, have been found to produce inhibition resulting in increased urinary excretion of heme precursors. Please see Chapters 4 and 5 for a more detailed discussion.

pathway is primarily localized in the endoplasmic reticulum and regulated by heme oxygenase.

2. *What is arsenic?*
   Arsenic is number 33 in the Periodic Table of Elements and may exist in oxidation states of 0, +/−3 and +5. It has been used for thousands of years as both a poison and a medicine (Fowler et al., 2015). It is also used in a number of industrial processes.

3. *Where is arsenic found?*
   Arsenic is found in the Earth's crust, in ore deposits with other elements such as copper and ground water flowing through arsenic containing rock formations (Fowler et al., 2015).

4. *How are humans exposed to arsenic?*
   Humans may be exposed to arsenic via inhalation of arsenic containing particles produced by smelting of ores such as copper or from coal burned in coal-fired power plants (Fowler et al., 2015). Arsenic exposure also occurs from intake of arsenic containing drinking water

(Fowler et al., 2015) and foodstuffs such as rice grown in paddies irrigated with arsenic containing water. In order to finish a complete and rigorous risk assessment for arsenic, it is essential to have solid analytical arsenic exposure data on the exposed populations with regard to the concentrations of arsenic in air, food, water, and urine so that the biomarker response may be placed in an exposure context. Ideally it would be best if all the exposure, metabolism, genetic polymorphism, and molecular biomarker endpoint data could be collected on the same persons so that individual risk assessments could be generated and sensitive subgroups are not overlooked. The findings on individuals could then be pooled on a group or population basis to meet practical regulatory needs which would then be informed on basis of MOA approaches.

5. *How is arsenic handled in the body?*
   Inorganic arsenic may undergo both oxidation/reduction reactions and methylation reactions in vivo resulting in the urinary excretion of inorganic methylated arsenical species in the urine (Fowler et al., 2015).

6. *What are the known clinical effects of arsenic exposure?*
   Arsenic is a broad spectrum agent which is capable of affecting a number of organ systems including the liver, lungs, kidneys, skin, and hematopoietic system (Fowler et al., 2015). It may produce both organ toxicity and cancer depending on dose, duration of exposure, and individual susceptibility (Fowler et al., 2015).

7. *What is a metabolomic biomarker and why is the heme pathway useful in this regard?*
   A metabolomic biomarker is a biologically generated molecule which is a component of metabolic pathway whose presence in a tissue or other matrix such as blood, urine, or feces is altered by drug or chemical exposures. It is reflective of the disruptive biological activity of drug or chemical of interest to that metabolic pathway and provides a translational bridge between exposure and effect at an early stage of toxicity. Chemical-induced alterations in the heme pathway enzyme activities and increased excretion of metabolic intermediates such as ALA and specific porphyrin isomers have been used as early and specific biomarkers of cellular toxicity (Fowler, 2009, 2012).

8. *What are the known mechanistic effects of arsenic on the heme pathway in experimental systems?*
   Inorganic arsenical exposures are have been shown to produce a porphyrinuria patterns in rodents (Woods and Fowler, 1978) and humans (Garcia-Vargas et al., 1994). The underlying mechanisms are related to inhibition of the enzymes uroporphrinogen decarboxylase (Woods and Fowler, 1987) and coporporhyrinogen oxidase (Woods and Fowler, 1987; Woods et al., 1981) and ferrochelase (Woods and Fowler, 1977).

9. *What is known about the mode of arsenical action on heme biosynthetic pathway enzymes?*
   The mode of action appears to be related to production of ROS generated from inhibition of mitochondrial respiration (Fowler et al., 1979).

10. *What are the specific porphyrins excreted in urine following arsenic exposure?*
    Uroporphyrins (8, 7, 6, 5-carboxyl) and 4-carboxyl coproporphyrin (Woods et al., 1981).

11. *What roles could arsenic-induced disturbances in the heme biosynthetic play in the development of arsenic-induced organ toxicity and cancer?*
    Arsenic-induced inhibition of heme biosynthesis and inhibition of mitochondrial respiration appear to be major factors in driving overt cell injury/cell death. The associated turnover of cells in combination with arsenic-induced oxidative stress from increased production of ROS and oxidative stress inducing porphyrins is a possible overall mechanism for cancer induction (Flora et al., 2007; Flora, 2011; Jomova et al., 2011).

12. *How can measurement of alterations in heme pathway enzymes and porphyrins support MOA risk assessments for arsenic-induced toxicity and carcinogenicity?*
    Based upon the previously mentioned known mechanisms for arsenic-induced effects on cellular heme metabolism and the roles of oxidative stress in cell injury/cell death and carcinogenesis, it is reasonable to conclude that the described arsenical disturbances in heme and porphyrin metabolism may be useful as early harbingers for arsenic-induced clinical toxicities and development of cancer. These effects may hence be of basic science value in MOA-based risk assessments strategies. This opinion is supported by similar arsenic-induced porphyrinuria effects and actual health outcomes in Mexican populations exposed to arsenic in drinking water (Garcia-Vargas et al., 1994; García-Vargas and Hernández-Zavala, 1996; Hernández-Zavala et al., 1999) and populations in China exposed to arsenic from burning of coal (Ng et al., 2005; Wang et al., 2002; Xie et al., 2001).

3   TRANSLATION OF MOLECULAR BIOMARKER DATA FOR SOCIETAL DECISION MAKING

3.1   Communicating With the Professional Risk Assessment Community

Once molecular biomarker data have been explained in readily understandable terms, this information must be accepted by the professional risk assessment community as valid and useful for helping to improve risk assessment judgments. In other words, does inclusion of biomarker-based MOA information strengthen the scientific case for a given risk assessment judgment? The answer to this question will rest in large part on whether a number of the points raised in prior chapters of this book have been addressed. If there are holes in the data presented or procedures followed, this will weaken the case and make it less likely that the molecular biomarker data will be considered as valuable. On the other hand, if well-curated molecular biomarker data (using the guideline suggestions discussed in earlier chapters of this book) are available to support a given risk assessment and not included, then ultimately such a risk assessment will be considered antiquated and perhaps inadequate. Clearly, there is much to be gained by inclusion of solid molecular biomarker data in any modern risk assessment. This concept is the basis for the MOA approach to risk assessment which is gaining credibility in countries and agencies.

3.2 Communicating With Societal Decision Makers

Once the professional risk assessment community has reviewed and agreed to the inclusion of molecular biomarker data in a given risk assessment, the data and conclusions based upon this information should be submitted to proper peer review which will usually include publication in a solid peer-reviewed scientific journal ideally with a high publication impact factor. If the outcome of the risk assessment is of great importance to a major regulatory decision, the data, publications, and conclusions may be forwarded to independent scientific review committees such as those of the EPA SAB or a committee of the NAS/NRC. These external peer-review activities are usually time consuming but necessary in order to give confidence to societal decision makers such as high level agency regulators and/or legislators, who must translate the scientific information into regulations or legislation based on the validity of the risk assessment conclusions. These types of decisions are not trivial and the process of moving from scientific judgments to societal actions is not taken lightly since there is usually a number of both expected and unintended consequences to any regulation or enacted piece of legislation.

In addition to formal regulatory or legislative acceptance of a given risk acceptance, there is another audience composed of the lay public and stakeholder who will be impacted either directly or indirectly by a regulation or piece of legislation based upon given risk assessment. If the risk assessment is based or supported in part by molecular biomarker data, then it is imperative that the attendant risk communication information provide a fundamental understanding of supporting molecular biomarker data in plain English for persons with limited technical backgrounds.

3.3 Communicating With the Lay Press and Lay Public

In addition to the previously mentioned important groups, it is very important to translate molecular biomarker into terms that may be appreciated by the lay press and lay public (Bennett and Waters, 2000; Castiel, 1999; Daughton, 2001;

Liotta et al., 2005). This aspect is not a small matter since these groups may have limited technical backgrounds but great interest in a particular health issue which has been discussed on the Internet by various interests with different opinions. This complicates matter since this situation may create confusion and make presentation of a new molecular biomarker test system more difficult to explain in presence of prior conflicting opinions. In other words, the putative biomarker must be explained both in terms of its intrinsic value as a new test but also how it helps resolve existing differences of opinion which impacts a particular risk assessment. This means that the risk communication in such a situation must be both clear, accurate, and provided, as noted earlier, in "plain English" so that the message is correctly understood and there is minimal chance for misunderstanding by either reporters or the readers of the press publication. In this regard, use of readily understood graphics to explain a putative biomarker may be particularly helpful to achieve a good communication between the scientific community and a lay audience.

3.3.1 Importance of Informative Graphic Presentations in Communicating via the Media

Given the technical complexity of molecular biomarkers, it is clear that effective communication of the data presented and explanations of the meaning of the data is enhanced by good graphics. In other words, the adage that "a picture is worth a thousand words" holds true here as well. This is particularly true in dealing with the lay press and television media where there is limited space and time in which to effectively communicate findings so that they are clearly understood (Feero et al., 2010; Haga et al., 2014; Hawk et al., 2008; Seals, 2013).

3.3.2 Social Media Presentations

Following up on the previous point is the recognition that social media communications are now a major vehicle by which many people, particularly those in younger generations research and receive information. It is hence important to give attention to how to communicate biomarker information via these information outlets (Seals, 2013; West, 2013). Many of the social media formats are, by necessity, truncated in length which presents a particular challenge to those wishing to communicate information about molecular biomarkers in an effective manner but this can be achieved by distillation of the message into plain English and use of carefully crafted graphics to tell the story.

4 SUMMARY AND CONCLUSIONS

Based upon the previous discussion, it is clear that communication of risk for adverse health effects based upon incorporation of molecular biomarkers into a risk assessment application offers both challenges and opportunities. On one hand, the challenges include the need to include an educational component in the communication to explain both the value of the putative biomarker on a mechanistic basis as well as how inclusion of biomarker data to the risk assessment adds value and credibility to the final assessment. On the other hand, inclusion of molecular biomarker data provide a scientific foundation for utilizing an MOA approach to the risk assessment which, in theory, offers greater sensitivity, specificity, and precision to the overall risk assessment. These aspects are of particular importance in dealing with situations where there are concerns about sensitive populations and mixtures of chemicals or drugs. These types of factors are of ever increasing importance in risk assessment practice for both chemicals and drugs.

References

Bennett, D.A., Waters, M.D., 2000. Applying biomarker research. Environ. Health Perspect. 108 (9), 907.

Castiel, L.D., 1999. Apocalypse... Now? Molecular epidemiology, predictive genetic tests, and social communication of genetic contents. Cadernos De Saúde Pública 15, S73–S89.

Cutts, M., 2013. Oxford Guide to Plain English. Oxford University Press, Oxford, UK.

Daughton, C.G., 2001. Emerging pollutants, and communicating the science of environmental chemistry and mass spectrometry: pharmaceuticals in the environment. J. Am. Soc. Mass Spectrom. 12 (10), 1067–1076.

Deng, G., Zheng, B., Zhai, C., Wang, J., Ng, J., 2007. Porphyrins as the early biomarkers for arsenic exposure of human. Huan jing ke xue 28 (5), 1147–1152.

Ellis, D., Barker, R., Potter, S., Pridgeon, C., 1993. Information audits, communication audits and information mapping: a review and survey. Int. J. Inform. Manag. 13 (2), 134–151.

Feero, W.G., Guttmacher, A.E., Feero, W.G., Guttmacher, A.E., Collins, F.S., 2010. Genomic medicine—an updated primer. N. Engl. J. Med. 362 (21), 2001–2011.

Flora, S.J., 2011. Arsenic-induced oxidative stress and its reversibility. Free Radic. Biol. Med. 51 (2), 257–281.

Flora, S., Bhadauria, S., Kannan, G., Singh, N., 2007. Arsenic induced oxidative stress and the role of antioxidant supplementation during chelation: a review. J. Environ. Biol. 28 (2), 333.

Fowler, B.A., 2009. Monitoring of human populations for early markers of cadmium toxicity: a review. Toxicol. Appl. Pharmacol. 238 (3), 294–300.

Fowler, B.A., 2012. Biomarkers in toxicology and risk assessment. EXS 101, 459–470.

Fowler, B.A., Chou, S.-H.S., Jones, R.L., Sullivan, Jr., D.W., Chen, C.-J., 2015. Arsenic. In: Nordberg, G.F., Fowler, B.A., Nordberg, M. (Eds.), Handbook on the Toxicology of Metals. fourth ed. Elsevier Publishers, Amsterdam, pp. 582–624.

Fowler, B.A., Woods, J.S., Schiller, C.M., 1979. Studies of hepatic mitochondrial structure and function: morphometric and biochemical evaluation of in vivo perturbation by arsenate. Lab. Invest. 41 (4), 313–320.

García-Vargas, G.G., Hernández-Zavala, A., 1996. Urinary porphyrins and heme biosynthetic enzyme activities measured by HPLC in arsenic toxicity. Biomed. Chromat. 10 (6), 278–284.

Garcia-Vargas, G.G., Del Razo, L.M., Cebrian, M.E., Albores, A., Ostrosky-Wegman, P., Montero, R., et al.,1994. Altered urinary porphyrin excretion in a human population chronically exposed to arsenic in Mexico. Hum. Exp. Toxicol. 13 (12), 839–847.

Guanci, G., 2009. Best practices for webinars. Creative Nursing 16 (3), 119–121.

Haga, S.B., Mills, R., Bosworth, H., 2014. Striking a balance in communicating pharmacogenetic test results: promoting comprehension and minimizing adverse psychological and behavioral response. Patient Edu. Couns. 97 (1), 10–15.

Hawk, E.T., Matrisian, L.M., Nelson, W.G., Dorfman, G.S., Stevens, L., Kwok, J., et al.,2008. The Translational Research Working Group developmental pathways: introduction and overview. Clin. Cancer Res. 14 (18), 5664–5671.

Hernández-Zavala, A., Del Razo, L.M., García-Vargas, G.G., Aguilar, C., Borja, V.H., Albores, A., Cebrián, M.E., 1999. Altered activity of heme biosynthesis pathway enzymes in individuals chronically exposed to arsenic in Mexico. Arch. Toxicol. 73 (2), 90–95.

Horn, R.E., 1974. Information mapping. Training Bus. Ind. 11 (3), 27–32.

Houghton, J.W., Pucar, M., Knox, C., 1996. Mapping information technology. Futures 28 (10), 903–917.

Jomova, K., Jenisova, Z., Feszterova, M., Baros, S., Liska, J., Hudecova, D., et al.,2011. Arsenic: toxicity, oxidative stress and human disease. J. Appl. Toxicol. 31 (2), 95–107.

Liotta, L.A., Lowenthal, M., Mehta, A., Conrads, T.P., Veenstra, T.D., Fishman, D.A., Petricoin, E.F., 2005. Importance of communication between producers and consumers of publicly available experimental data. J. Nat. Cancer Inst. 97 (4), 310–314.

McCarthy, S., Saxby, L., Thomas, M., Weertz, S., 2012. Connecting Through Webinars: A CRLA Handbook for the Use of Webinars in Professional Development. College Reading and Learning Association Professional Development Committee. Recuperado a parti r de. Available from: http://www.crla.net/ProfDev/Connecting%20through%20Webinars%20 CRLA%20Handbook.pdf

Mohorovičić, S., Lasic-Lazic, J., Strčić, V., 2011. Webinars in higher education. Paper presented at the MIPRO, Proceedings of the 34th International Convention.

Molay, K., 2009. Best Practices for Webinars.

Ng, J.C., Wang, J.P., Zheng, B., Zhai, C., Maddalena, R., Liu, F., Moore, M.R., 2005. Urinary porphyrins as biomarkers for arsenic exposure among susceptible populations in Guizhou province. China Toxicol. Appl. Pharmacol. 206 (2), 176–184.

Seals, D.R., 2013. Translational physiology: from molecules to public health. J. Physiol. 591 (14), 3457–3469.

Thrush, E.A., 2001. Plain English? A study of plain English vocabulary and international audiences. Tech. Commun. 48 (3), 289–296.

Wang, J., Qi, L., Zheng, B., Liu, F., Moore, M., Ng, J., 2002. Porphyrins as early biomarkers for arsenic exposure in animals and humans. Cell. Mol. Biol. 48 (8), 835–843.

West, H.J., 2013. Practicing in partnership with Dr. Google: the growing effect of social media in oncology practice and research. Oncologist 18 (7), 780–782.

Woods, J.S., Fowler, B.A., 1977. Renal porphyrinuria during chronic methyl mercury exposure. J. Lab. Clin. Med. 90 (2), 266–272.

Woods, J.S., Fowler, B.A., 1978. Altered regulation of mammalian hepatic heme biosynthesis and urinary porphyrin excretion during prolonged exposure to sodium arsenate. Toxicol. Appl. Pharmacol. 43 (2), 361–371.

Woods, J.S., Fowler, B.A., 1987. Metal alteration of uroporphyrinogen decarboxylase and coproporphyrinogen. Ann. NY Acad. Sci. 514, 55–64.

Woods, J.S., Kardish, R., Fowler, B.A., 1981. Studies on the action of porphyrinogenic trace metals on the activity of hepatic uroporphyrinogen decarboxylase. Biochem. Biophys. Res. Commun. 103 (1), 264–271.

Xie, Y., Kondo, M., Koga, H., Miyamoto, H., Chiba, M., 2001. Urinary porphyrins in patients with endemic chronic arsenic poisoning caused by burning coal in China. Environ. Health Prevent. Med. 5 (4), 180–185.

Zaragoza-Anderson, K.M., 2008. Online webinars for continuing medical education: an effective method of live distance learning. Int. J. Instruct. Tech. Dist. Learn. 2 (8), 7–14.

CHAPTER 10

Future Research Directions

1 FURTHER VALIDATION OF BIOLOGICAL MARKERS FOR HUMANS AND BARRIERS TO ACCEPTANCE INTO RISK ACCEPTANCE PRACTICE

As noted in prior chapters of this book, biomarker validation through linkage to human health endpoints are key components for acceptance of molecular biomarkers as significant tools in risk assessment practice. There are several needed levels to this translational process which align with the levels of biological organization. This chapter will identify current areas of needed translational information for which data are available and suggest ways in which this information base may be further developed to move the field of biomarker-based risk assessment forward. It should be noted that the areas identified in following sections could be readily developed with relatively little effort and expense which will also be important considerations in terms of their application. This list is also not inclusive but provided to demonstrate the possibilities for using biomarkers to improve risk assessments in the immediate future.

1. Linkages to other health endpoints and toxicological processes such as adverse outcome pathways which provide information on the prognostic significance of the putative biomarker in humans at the cellular level:Establishment of this linkage ideally requires integration of morphological and biochemical data in target cell populations with the measured biomarker data from human cells. This is a difficult but not impossible task if one makes use of recent advantages in cell biology to study relevant human cell line systems in a dose-response context. Computational models can then be used to extrapolate from the in vitro data to intact organ systems. This approach should generate useful estimates of the impact of a given chemical or mixture of chemicals on a biomarker endpoint in a target cell population. Risk assessment conclusions can be justified with actual mechanistic scientific data using this mode of action (MOA) approach. The precision of this approach will doubtlessly improve over time as

Molecular Biological Markers for Toxicology and Risk Assessment. http://dx.doi.org/10.1016/B978-0-12-809589-8.00010-X

understanding of the linkages between the biomarker of interest and adverse outcome pathways continues to grow. This will lead to further refinements in the MOA approach and increased confidence in projected risk assessments.

2. Linkages to other clinical endpoints and human health outcomes using epidemiological data: *The Bradford Hill Criteria for Principles of Causality*. In order for a putative biomarker to gain broad acceptance and be incorporated into risk assessment practice, it must be validated by established criteria. This issue has been appreciated for some time in the field of clinical research. The most well-known set of criteria for this purpose are those articulated by Bradford Hill (1965) and which have been extended by others to different fields in which experimental data on humans are seldom, if ever, available. The Bradford Hill criteria are shown in Table 10.1.

3. Biological marker–based risk assessments and application of the Bradford Hill Criteria: The criteria for establishing causality articulated by Bradford Hill for validation of conclusions using standard epidemiological or clinical endpoints are appropriate for validating molecular biomarker endpoints. The potential value of incorporating these biomarker endpoints over standard clinical endpoints rests with improvements in sensitivity and specificity of biological responses that permit *early detection* of toxicity *prior* to an adverse outcome event rather than *after* it has occurred. Use of molecular biomarker endpoints are hence potentially powerful tools for *prevention* of drug/chemical-induced toxicity. This would be of particular value for sensitive subpopulations at special risk for toxicity related to, for example, age, gender, race, nutritional status, and genetic inheritance. The ability to identify these modifying factors and incorporate their impact into an overall risk assessment for a given drug or chemical is one of the great potential advantages of applying molecular biomarker to risk assessments. Ideally, application of these sensitive tools could ultimately lead to individual-based risk assessments for drug/chemical toxicity. This approach could directly detect individuals in sensitive

---

Table 10.1 Bradford Hill's Principles of Causality in Clinical Research

- Strength of the association
- Specificity of the association
- Evidence of a dose-response relationship
- Biological plausibility of the hypothesis
- Coherence of the evidence
- Temporality
- Consistency of results across studies

Source: Hill (1965).

subpopulations. This approach is consonant with ongoing efforts to expand personalized precision medicine and reduce costs of medical treatment. The same issues would hold true for the expanded use of molecular biomarker–based diagnostics in early detection of toxicity and prevention of adverse clinical outcomes. Clearly these tools are compatible with ongoing personalized medicine approaches and could be incorporated into routine physical examinations at minimal additional cost. If incorporation of validated molecular biomarkers were to become a component of routine physical examinations, this could potentially prevent a number of major illnesses and cancer by early detection when effective intervention is possible. The health benefits of such early detection would likely cover the additional costs of the biomarker tests for the physical examination and provide additional data on the efficacy of a given biomarker in humans for validation purposes. It should be noted that this approach would be somewhat analogous to the current FDA postapproval monitoring of drugs in clinical use which adds to the information database on safety and efficacy from experience over time. It is also clear that such an approach in combination with the ongoing evolution of personalized precision medicine will generate enormous quantities of data which will require effective methods of analysis in order to be of value for risk assessment purposes. As discussed later, computational toxicology and eventually application of artificial intelligence (AI) systems will be essential to the analysis of the extremely large data sets generated by these or similar large scale approaches.

2   APPLICATION OF ARTIFICIAL INTELLIGENCE COMPUTER PROGRAMS FOR INTEGRATING DIVERSE DATA SETS AND FACILITATING RISK ASSESSMENTS

With molecular biomarkers, the field of AI has been rapidly evolving (Jordan and Mitchell, 2015) due, in part, to the need to handle the large quantities of data generated by robotics-driven biomarker analysis systems (Ilyin et al., 2004; Ross et al., 2005). Large data sets produced as a result of interdisciplinary studies from collaborating laboratories is another evolving area that could potentially generate enormous quantities of data resulting in the need for "big data" analysis approaches. This area becomes also more complex if one wishes to integrate molecular biomarker data with other types of relevant information such as data from exposure analyses and standard clinical tests. On the one hand, such a global approach would permit the development of more comprehensive risk assessments and precise risk assessments. On the other hand, expanded computational methods to handle the large and diverse data sets are required. This is as major undertaking. However, the use of AI systems to an even greater extent in the future seems a realistic possibility, given that these systems are already being used for risk assessments in a variety of contexts. (Chen et al., 2014; Ginex et al., 2014; Javadikasgari et al., 2014; Jia et al., 2014; Li et al., 2014; Liu et al., 2013; Singh et al., 2015; Zhang et al., 2013). As these AI tools continue to evolve, they will be applied to ever more complex and difficult toxicology issues such as chemical mixtures and incorporation of individual genetic data into risk assessments.

3   CALCULATION OF ACCEPTABLE EXPOSURE LEVELS FOR CHEMICALS ON AN INDIVIDUAL OR MIXTURE BASIS

One of the most challenging problems faced by the field of risk assessment is how to conduct accurate and credible evaluations of chemical risk for new or unstudied drugs/chemicals such as nanomaterials (Linkov et al., 2007; Savolainen et al., 2010) for which there are few or no available data and for mixtures of drugs/chemicals where interactions may occur. For mixture situations, it is important to delineate whether possible interactions are additive, synergistic, or antagonistic in nature since these effects may greatly alter the accuracy of a predicted dose-response relationship and hence ultimately overall risk assessment conclusions. This general problem has been appreciated for a number of years by Federal agencies such as ATSDR (2004) and individual groups of investigators (Conner et al., 1993; Demchuk et al., 2011; Fowler, 2012; Fowler et al., 2004; Madden and Fowler, 2000; Savolainen et al., 2010; von Stackelberg et al., 2015; Whittaker et al., 2011). As with the other needed areas of research, chemical mixtures is another problem area which is complex and growing but not impossible to address if the evolving tools of computational toxicology (Fowler, 2013) are utilized effectively.

4   INCORPORATION OF INDIVIDUAL GENOTYPES INTO BIOLOGICAL MARKER-BASED RISK ASSESSMENTS

Another important and relatively underutilized area of available data concerns the incorporation of human genotypic data from large databases, such as NHANES, into risk assessment practice. This information is available for a number of important genes (Chang et al., 2009; Janssens et al., 2011) and could be expanded through further genotypic analysis of archived samples or new NHANES sampling. This database has been used on a limited basis for assessing the relative risk of hypertension from lead exposure on the basis of the ALAD 1 and 2 alleles (Scinicariello et al., 2010). The database could be used to improve risk assessments for a number of other chemicals known to be handled by metabolic systems with multiple allelic genotypes. The incorporation of available genetic information on the general US population could identify sensitive subpopulations at special risk for chemical-induced toxicity.

5 GETTING ON THE RISING ROAD—SOME SUGGESTIONS

In summary, there are a number of advances in areas related to molecular biomarkers that could be readily developed to improve risk assessment practice. These activities are occurring in both the public and private sectors; however, the translation of available information into practical usage is still at an early stage of development. Some suggested basic steps needed are as follows:

1. Data-mining of existing public and private (where available) for both hypothesis generation and risk assessment evaluations. The collective scientific community has information that could be analyzed for baseline risk assessments and identification of populations at special risk for toxicity. The NHANES data set is a classic example of this approach. These studies could be easily expanded to include inclusion of molecular biomarkers to produce more precise and MOA-based risk assessments.
2. Utilize automated biomarker test systems linked to computer management and data analysis systems. Such systems currently exist and are rapidly evolving.
3. Increase the use of AI systems for data analysis and handling of large interdisciplinary data sets. The use of AI for risk assessment purposes will become essential given the enormous quantities of complex analytical, biochemical, and clinical data that will be included in future risk assessments.
4. Develop a clear and immediate communicated vision that the road in front of us is rapidly rising in the direction of personalized medicine and hence the call for individual-based risk assessments is inevitable. The trend has already begun. As noted elsewhere in this book different risk characteristics, such as age, gender, nutritional status, race, and genetic traits play important roles in determining populations at special risk for toxicity. The long-term value of this approach is that it will increase the prevention of chemical-induced diseases and result in major savings in healthcare costs by virtue of decreasing adverse health effects.
5. An essential element for the effective translation and acceptance of molecular biomarkers into risk assessment practice is *effective communication* of how these complex modern tools improve risk assessments and better protect the public health and reduce healthcare costs. As with any new thing, the field of molecular biomarkers for risk assessment must be effectively communicated in order to be understood and ultimately achieve acceptance into common risk assessment practice.

References

ATSDR, 2004. Guidance Manual for the Assessment of Joint Toxic Action of Chemical Mixtures. U.S. Department of Health and Human Services, Atlanta, GA, 63pp.

Chang, M.H., Lindegren, M.L., Butler, M.A., Chanock, S.J., Dowling, N.F., Gallagher, M., et al., 2009. Prevalence in the United States of selected candidate gene variants: Third National Health and Nutrition Examination Survey, 1991–1994. Am. J. Epidemiol. 169 (1), 54–66.

Chen, Y., Cheng, F., Sun, L., Li, W., Liu, G., Tang, Y., 2014. Computational models to predict endocrine-disrupting chemical binding with androgen or oestrogen receptors. Ecotoxicol. Environ. Saf. 110, 280–287.

Conner, E.A., Yamauchi, H., Fowler, B.A., Akkerman, M., 1993. Biological indicators for monitoring exposure/toxicity from III–V semiconductors. J. Expo. Anal. Environ. Epidemiol. 3 (4), 431–440.

Demchuk, E., Ruiz, P., Chou, S., Fowler, B.A., 2011. SAR/QSAR methods in public health practice. Toxicol. Appl. Pharmacol. 254 (2), 192–197.

Fowler, B.A., 2012. Biomarkers in toxicology and risk assessment. EXS 101, 459–470.

Fowler, B.A. (Ed.), 2013. Computational Toxicology: Applications for Risk Assessment. Elsevier Publishers, USA, 258.

Fowler, B.A., Whittaker, M.H., Lipsky, M., Wang, G., Chen, X.Q., 2004. Oxidative stress induced by lead, cadmium and arsenic mixtures: 30-day, 90-day, and 180-day drinking water studies in rats: an overview. Biometals 17 (5), 567–568.

Ginex, T., Spyrakis, F., Cozzini, P., 2014. FADB: a food additive molecular database for in silico screening in food toxicology. Food Addit. Contam. Part A Chem. Anal. Control Expo. Risk Assess. 31 (5), 792–798.

Hill, A.B., 1965. The environment and disease: association or causation? Proc. Royal Soc. Med. 58 (5), 295–300.

Ilyin, S.E., Belkowski, S.M., Plata-Salamán, C.R., 2004. Biomarker discovery and validation: technologies and integrative approaches. Trends Biotechnol. 22 (8), 411–416.

Janssens, A.C., Ioannidis, J.P., van Duijn, C.M., Little, J., Khoury, M.J., 2011. Strengthening the reporting of Genetic RIsk Prediction Studies: the GRIPS Statement. PLoS Med. 8 (3), e1000420.

Javadikasgari, H., Ghavidel, A.A., Gholampour, M., 2014. Genetic fuzzy system for mortality risk assessment in cardiac surgery. J. Med. Syst. 38 (12), 155.

Jia, X., Miao, Z., Li, W., Zhang, L., Feng, C., He, Y., et al., 2014. Cancer-risk module identification and module-based disease risk evaluation: a case study on lung cancer. PLoS One 9 (3), e92395.

Jordan, M.I., Mitchell, T.M., 2015. Machine learning: trends, perspectives, and prospects. Science 349 (6245), 255–260.

Li, X., Chen, L., Cheng, F., Wu, Z., Bian, H., Xu, C., et al., 2014. In silico prediction of chemical acute oral toxicity using multi-classification methods. J. Chem. Inf. Model 54 (4), 1061–1069.

Linkov, I., Satterstrom, F.K., Steevens, J., Ferguson, E., Pleus, R., 2007. Multi-criteria decision analysis and environmental risk assessment for nanomaterials. J. Nanopart. Res. 9 (4), 543–554.

Liu, J., Page, D., Nassif, H., Shavlik, J., Peissig, P., McCarty, C., et al., 2013. Genetic variants improve breast cancer risk prediction on mammograms. AMIA Annu. Symp. Proc., 2013, 876–885.

Madden, E.F., Fowler, B.A., 2000. Mechanisms of nephrotoxicity from metal combinations: a review. Drug Chem. Toxicol. 23 (1), 1–12.

Ross, J.S., Symmans, W.F., Pusztai, L., Hortobagyi, G.N., 2005. Pharmacogenomics and clinical biomarkers in drug discovery and development. Am. J. Clin. Pathol. 124 (Suppl. 1), S29–S41.

Savolainen, K., Alenius, H., Norppa, H., Pylkkänen, L., Tuomi, T., Kasper, G., 2010. Risk assessment of engineered nanomaterials and nanotechnologies—a review. Toxicology 269 (2–3), 92–104.

Scinicariello, F., Yesupriya, A., Chang, M.H., Fowler, B.A., 2010. Modification by ALAD of the association between blood lead and blood pressure in the U.S. population: results from the Third National Health and Nutrition Examination Survey. Environ. Health Perspect. 118 (2), 259–264.

Singh, A., Nadkarni, G., Gottesman, O., Ellis, S.B., Bottinger, E.P., Guttag, J.V., 2015. Incorporating temporal EHR data in predictive models for risk stratification of renal function deterioration. J. Biomed. Inform. 53, 220–228.

von Stackelberg, K., Guzy, E., Chu, T., Claus Henn, B., 2015. Exposure to mixtures of metals and neurodevelopmental outcomes: a multidisciplinary review using an adverse outcome pathway framework. Risk Anal. 35 (6), 971–1016.

Whittaker, M.H., Wang, G., Chen, X.Q., Lipsky, M., Smith, D., Gwiazda, R., Fowler, B.A., 2011. Exposure to Pb, Cd, and As mixtures potentiates the production of oxidative stress precursors: 30-day, 90-day, and 180-day drinking water studies in rats. Toxicol. Appl. Pharmacol. 254 (2), 154–166.

Zhang, Q., Yang, X., Zhang, Y., Zhong, M., 2013. Risk assessment of groundwater contamination: a multilevel fuzzy comprehensive evaluation approach based on DRASTIC model. ScientificWorldJournal 2013, 610390.

Index

Printed in the United States
By Bookmasters